D1106515

NO LONGER PROPERTY OF
GLENDALE LIBRARY,
ARTS & CULTURE DEPT.

05/06
2

GLENDALE PUBLIC LIBRARY
222 East Harvard St.
Glendale, CA 91205

BEGINNER'S GUIDE TO

paper**making**

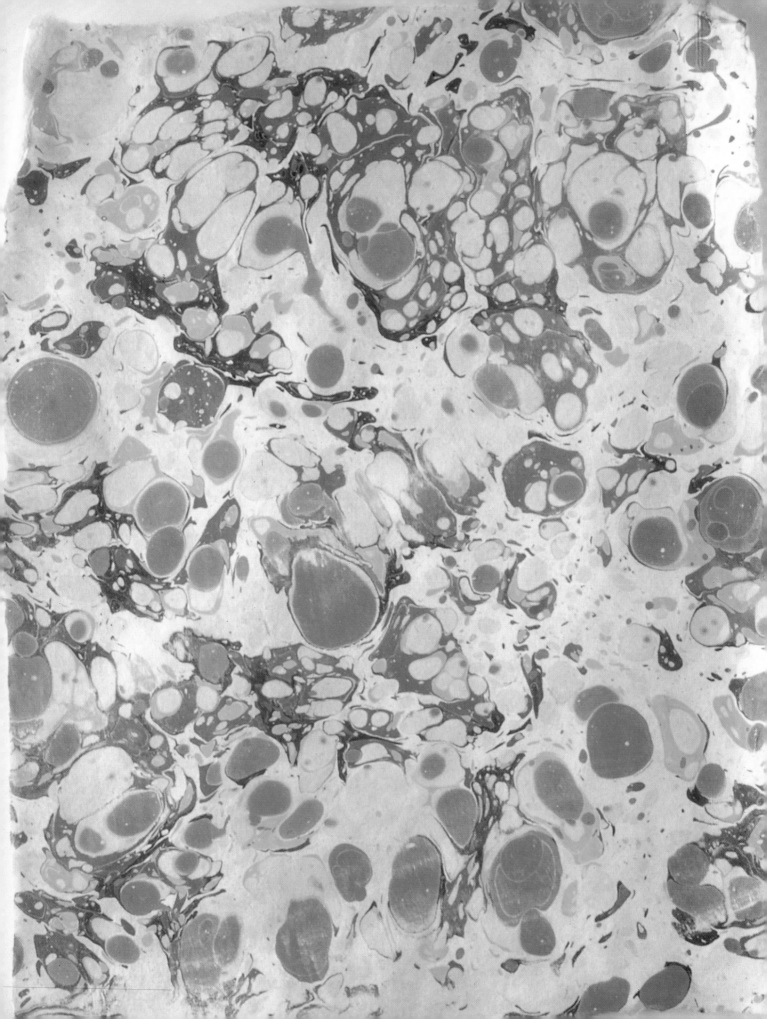

BEGINNER'S GUIDE TO

papermaking

heidi reimer-epp & mary reimer

676.22
REI

Sterling Publishing Co., Inc.
New York

Creative director: Sarah King
Editor: Judith Millidge
Project editor: Anna Southgate
Designer: Axis Design Editions

Library of Congress Cataloging-in-Publication Data Available

10   9   8   7   6   5   4   3   2   1

Published in 2003 by Sterling Publishing Co., Inc.
387 Park Avenue South, New York, N.Y. 10016

This book was designed and produced by
D&S Books Ltd
Kerswell, Parkham Ash, Bideford, Devon, EX39 5PR

© 2003 D&S Books Ltd

Distributed in Canada by Sterling Publishing
C/o Canadian Manda Group,
One Atlantic Avenue, Suite 105
Toronto, Ontario, Canada M6K 3E7

Every effort has been made to ensure that all the information in this book is accurate.
however, due to differing conditions, tools, and individual skills, the publisher cannot
be responsible for any injuries, losses, and other damages which may result from the
use of the information in this book.

Printed in Hong Kong

All rights reserved.

Sterling ISBN 0-8069-9394-4

contents

introduction to the history of papermaking

When you think about it, paper is all about communication! Look around you. You are probably within reach of a pen or pencil, and some form of notepad, message board, or sticky notes. Even in this technology-driven time, we rely on bits of paper on which we scribble phone numbers, addresses, grocery lists, and to-do lists. Even those of us who are sufficiently technologically advanced to enter all of the above straight into our personal organizers or computers must accept restaurant checks, store receipts, bank statements, and so on, in paper form. Our pocketbooks are crammed with significant scraps of paper that hold essential bits of information. How many business deals have been worked out on a paper napkin, over lunch in a restaurant? Framed documents and credentials adorn the walls of many offices and boardrooms—all printed on fine paper and bearing official seals and signatures. Certainly, paper has also been used for less noble, but, nevertheless, essential functions. Where would we be without facial tissues and toilet paper, for example? And consider the newspaper and magazine industries—we know that we can easily read online, so why bother with the print form at all?

The sense of touch, that's why! We like to hold, to feel the roughness of a newspaper, the glossiness of the latest journal, the solid comfort of a good book. We like the smell of the paper, the glue, the oldness or newness of it. We even like to see the signs of someone else's journey through an old book—the soft edges, a thumbprint, a tear stain and even—horrors—a bent corner! Yes, paper is here to stay. Too many people have worked too hard to develop this remarkable bonding of fibers to let it pass into history, taking along with it the ultimate love letter, the final novel, and the last will and testament, carefully rolled and tied with a ribbon. So come along now, and join in on a journey that passes through time and space, tracing the paper path from its early beginnings through to its future.

the origins of paper

To begin our discovery of the origins of paper, it is important to understand exactly what constitutes paper. Mostly seen as a final product, as something other than paper (money, boxes, envelopes, invitations, tickets, etc.), paper itself is made up of the bonding of plant fibers.

In his classic work entitled *Papermaking in the Classroom*, Dard Hunter gives this definition of paper: "To be classed as true paper, the thin sheets must be made from fiber that has been macerated until each individual filament is a separate unit; the fibers intermixed with water, and by the use of a sieve-like screen, the fibers are lifted from the water in the form of a thin stratum, the water draining through the small openings of the screen, leaving a sheet of matted fiber upon the screen's surface. This thin layer of intertwined fiber is paper."

By definition then, the material from which we derive the name "paper" is not actually paper at all! papyrus, or *Cyperus papyrus*, is the marsh grass that grew along the banks of the Nile River in Egypt. Somewhere between 2000 and 3000 BC, Egyptians began the process of cutting thin strips from the stem of these grasses and placing them in the water to soften. These strips were placed side by side to form a layer, and then a second layer, with the strips running in the opposite direction to those in the first layer, was placed on top. This double-layered mat was then pounded until it formed a thin sheet, which was allowed to dry in the sun. These sheets were light, easily transported, and provided a good writing surface that could be rolled up and carried with ease. They soon became used for recording important information—essential records, documents, and early religious texts, in the Egyptian, Greek, and Roman worlds.

Previous to the use of papyrus, people expressed their creativity and recorded events of daily life in a variety of ways— stone carvings, cave drawings, and shapes chiseled into stone are just a few of these. Moses must have had quite a job carving the 10 Commandments into the stone tablets!

These effective but cumbersome methods gave way to clay tablets, wax-coated wood carvings, and, eventually, metal. Parchment and vellum are two early writing surfaces, which were made from animal skins. This process involved skinning the animal and scraping all the hair and fat from the skin. The skin was then carefully stretched and treated to produce a fine writing surface.

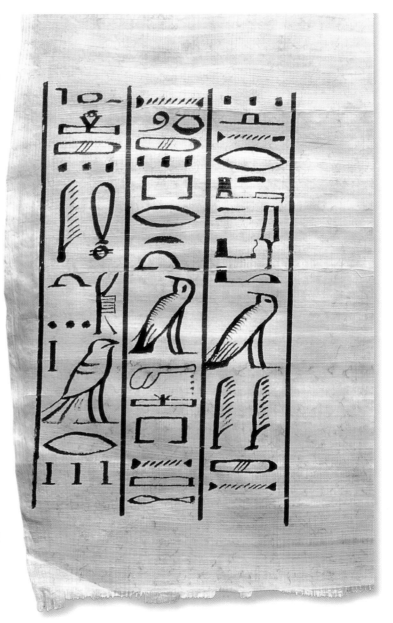

Papyrus is made from the pith of reeds grown in Egypt.

papermaking in china

None of the systems described above provided a convenient or cost-effective writing surface, and so the search continued. In China, writings on silk cloth have been discovered in the tombs of the former Han Dynasty (207 BC–AD 9). An interesting story tells of Ts'ai Lun, a peasant serving in the court of the Emperor Ho-Ti. The Queen, Dun-Shi, is said to have ordered him to create a writing

written language

The first written language—which used a system of wedge-shaped symbols or "cuneiform"—is attributed to the ancient Sumerians. The shapes they employed were scratched onto clay tablets and sometimes preserved inside clay tubes.

material, which would be cheaper than the silk paper used, to record the classic writings that the Queen loved. For many years he worked and experimented, barely taking time to eat or sleep. He used a combination of mulberry bark, hemp and linen remnants, and bits of fishing nets, which he soaked in water before beating the mixture until the fibers separated. Then he mixed the fibers with water in a large vat, and, using a cloth screen mold, scooped up a layer of fiber. The results, when dry, were the first sheets of paper! These strong, flexible sheets provided a fine writing surface, and Ts'ai Lun was given a place of distinction at court. His paper was named "Ts'ai Ko-Shi" or "Distinguished Ts'ai's Paper". Unfortunately, some years later, Ts'ai Lun fell into

Reconstruction of cuneiform lettering on clay tablet.

disgrace and, ashamed and dishonored, swallowed poison and died.

The knowledge of papermaking was a carefully guarded secret until the 3rd century, when it began its journey into other countries. First it traveled to Tibet and Vietnam, then, in the 4th century, it made its way to Korea, and, in the 6th century, Japan. Many fascinating tales and legends accompany the spread of papermaking throughout the world.

the spread of papermaking

Buddhist monks brought the technique from Korea to Japan, where the Emperor ordered the setting up of a paper production system. During the 8th century the Empress Shotuka is said to have ordered a massive project, in which a million prayers were to be printed on individual sheets of paper, each one mounted in its own pagoda. This project helped to establish Japan as a producer of high-quality paper, a reputation that continues to this day.

the journey westward

By AD 750, Arab armies were returning home from wars fought in central Asia. In AD 751, the Arab forces defeated the Chinese army at Samarkand, taking those who survived as prisoners. Among them were several skilled papermakers who were immediately put to work making paper out of plant fibers. Later, flax fiber and linen rags became sources for the Arabian papermill in Baghdad. Papermaking gradually spread throughout the Muslim world, to Damascus, to Egypt, and, eventually, to Europe with the invasion of Spain and Portugal by the Moors (who brought the knowledge with them). By the 9th century, paper was preferred to papyrus and parchment as a writing surface.

papermaking hits europe! enter – the machine!

The first European papermills were established in Spain in 1144, in France in 1348, in Fabriano, Italy in 1260, in Nuremburg, Germany in 1390 and in Hertford, England in 1490. The primary use of paper in the West was, up until this time, for religious purposes. With the Renaissance came increased literacy, and with it a demand for, and the availability of, a broader range of reading material. The invention of a movable typeface, by John Gutenburg, and the printing of the famous Gutenberg Bible in 1456, helped to spread Christianity, and also increased the importance of paper in mass communication. Business and commerce required paper for documents, certificates, and receipt books, as well as paper for daily use in wrapping and packaging.

Printing technology developed to meet the growing need for paper, which in turn resulted in a further increase in the demand for paper. Cotton and linen rags, as well as hemp fibers from rope and sailcloth, formed the raw materials for the pulp used in European papermaking. Water-driven stamping mills were invented to separate the fibers, a process that had previously been carried out by hand. Papermaking reached Holland in the 17th century, but stampers proved impossible to use there, as the flat land did not provide sufficient current to drive the waterwheel. In 1680, an enterprising Dutchman invented a beater, which ran on wind power. The Hollander beater greatly reduced the amount of time needed to turn rags into fibers. The large tub contained a revolving roller with cutting blades, which ground the rags against a stone plate.

The growth of efficiency in papermaking machines, however, resulted in a scarcity of rags. Several strategies were developed to deal with this shortage. In Germany and in England, dead bodies were no longer wrapped in cotton or linen cloths, but in wool instead. In America, aggressive advertizing campaigns urged citizens to contribute old clothing, for which they would receive some payment. One group of Americans came up with an extremely creative solution to the problem: they discovered that Egypt had a large supply of ancient mummies, each one of which was wrapped in several yards of linen. As a result, importing mummies became a great source of linen. This lasted until an outbreak of cholera developed among paper workers, caused by infected linen wrappings. The quest for a plentiful and inexpensive pulp source continued!

wood pulp

The use of wood pulp in papermaking began in the early 18th century, the result of observations by French naturalist, René de Reaumur. Reaumur noticed that the paper-like nests of Canadian wasps were formed from wood fibers, which the wasps chewed into a powder and mixed with an adhesive, produced within their bodies, to form a tough, water-resistant, multi-layered nest. From

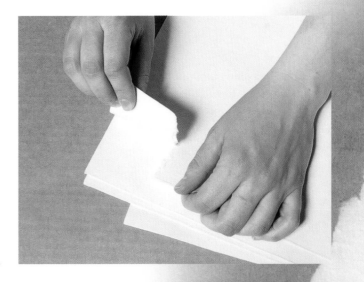

this he concluded that it must be possible to make paper by using the fibers from certain woods in place of rags. Although Reaumur had stumbled upon the answer to the question, and despite the fact that his work was both applauded and published, nothing was done to implement his ideas.

In 1839, over a hundred years after René de Reaumur's discovery, a young Canadian, Charles Fenerty, succeeded in making a machine to turn wood into pulp. Unfortunately, his years of work resulted in the production of only one sheet of paper!

Meanwhile, in the late 18th century, Frenchman Nicholas-Louis Robert produced a crude machine for grinding wood, which became the prototype for modern papermaking machines. This invention, which consisted of a sheet of wire mesh with squeeze rollers at one end, was capable of producing a seamless length of paper. Robert's intention had been to use the new machine to replace his workers, with whom he was constantly quarreling, but, owing to the turmoil of the French Revolution, he was unable to make any progress. A few years later, he moved to England, and joined forces with the English Fourdrinier brothers to develop a "new and improved" version of his

Today, paper is made from many materials, such as wood-pulp, flax fiber and hemp.

machine.

The pulp from this new machine produced thin, cheap paper, which served to revolutionize the communication industry in England by being used to print newspapers. Soon, Europe and America alike were producing massive volumes of paper for books, newspapers, money, packaging, and a seemingly endless list of purposes.[00]

what's happening today?

By the beginning of the American Civil War, only a few hand-papermaking mills remained, the art being practiced by just a handful of artists and craftspeople. In some areas of Asia, however, the art of making of paper by hand has continued to this day. In the West, a revival in handmade paper began in both Canada and the US during the 1960's and 1970's, largely brought about by artists creating papers for their own use in painting and printmaking. The substitution of wood pulp with "tree-free" fibers such as cotton, hemp, and abaca holds enormous appeal in a culture desperately trying to preserve forests for future generations.

Papermakers in the West are looking, once again, at the papermaking practices of the East, where indigenous plants are used to produce paper in the tradition of the past. Recycling, and the use of alternative fibers are the incentives behind many experimental papers currently being produced worldwide. It is estimated that, in the West, the average household throws away as much as 7lbs of packaging every week. Much of that paper could be recycled and, indeed, many hand papermakers are producing original papers from materials from their own recycling boxes.

Paper artists around the world are creating wonderful works of art, from the totally whimsical, such as Paul Johnson's 3D "Flying House", to the deliciously elegant "Yellow Petals over Steel Armature" by Helen Hiebert. The resurgence of interest in using handmade paper for writing letters and sending special invitations has stimulated the development of handbound books with beautiful covers made from fabric, leather, or even tree bark. Writing implements continue to evolve, to include fabulous colored inks, high-tech pens and pencils, as well as quill pens, and old-fashioned dipping pens. Where doomsayers once predicted the demise of writing and writing arts and skills, today's handmade papers and products are available in an endless variety and a constantly growing array.

Whether you are making paper for a specific purpose—say wedding invitations and book pages—or whether you are looking forward to producing one-of-a-kind sheets of paper purely for the joy of it, there are some essential pieces of equipment, and basic tools and materials that will help to ensure successful papermaking. As in any new undertaking, it

1

PAPERMAKING
getting started

is best to assemble everything you need before you begin – rather like baking a cake! If you are purchasing your mold and deckle (see Tools, pages 14–16), be sure to choose a vat size that will accommodate the dimensions of the model you have chosen.

The main factor is to set up a home papermill that suits your space and your needs. It is easy to increase size and production, but always begin with a setup that works well for you.

tools

Molds and deckles are the most important pieces of papermaking equipment. It is here that the pulp first makes contact with the form that shapes it, and this is the equipment that the papermaker treats with the greatest respect and care. There are many different mold constructions used in papermaking, and each requires a technique that is unique to that particular type of mold, but the principle of applying a layer of pulp fibers to a surface of woven material, through which the water can drain, is the same.

Early Chinese papermakers used a piece of coarsely woven cloth stretched over a four-sided bamboo frame. The pulp was scooped or poured directly onto the partially submerged screen, while the frame floated in a pool of water. It was then left to stand in the open air, to allow drying to occur. Once dry, the sheet was peeled away from the frame, which was then re-used. This process is still employed today in Nepal and other parts of the world.

Japanese papermakers use a flexible bamboo mold with a rigid frame, called a "su-geta". The technique of sheet formation involves continuous motion, which allows the fibers to align. Repetition of the dipping process builds up many thin layers into one sheet.

In Western papermaking, the "wove" mold is constructed using a frame, across which is

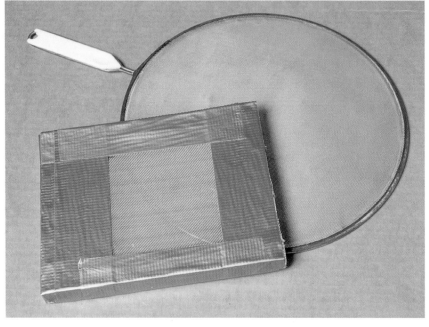

Assorted molds and deckles.

stretched a woven mesh screening (either wire or fabric). Pulp is scooped onto the mesh, allowing the water to drain through. A frame, called the "deckle", is placed on top of the mesh surface before sheet formation, in order to define the outside edges of the sheet.

Unlike Japanese paper, in which the fibers are aligned, in the wove method, the pulp fibers arrange themselves randomly on the mold. The"laid" mold is a second style of mold construction for this style of papermaking: it involves wires placed lengthwise and held in place by smaller wires running across in the other direction. Papers can be identified according to the pattern of the mold imprinted upon the finished sheet.

Home papermakers can use a variety of materials to construct a mold. Picture frames with mesh fabric stretched tightly across them are one possibility, while "spatter screens" from the kitchen allow you to form round sheets of paper. Build your own frames from wood with metal or fabric screening—heat-shrinking polypropylene is great for this, because it holds its shape well and can be tightened by heating with a blow-dryer if it starts to sag.

To mask off a section for smaller sheet formation, use

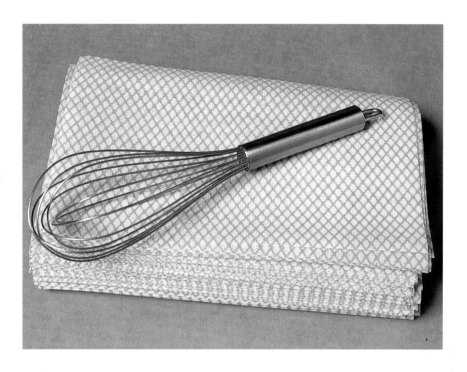

Couching cloths.

layers of masking tape, leaving the desired size and shape uncovered. This technique will be explored later in the book (see page 82).

Above all, always look after your mold and deckle sets, and take great pains to keep them clean and dry. After a papermaking session, carefully wash and dry all equipment and place on a rack to dry. A square of plastic ceiling grid makes a good drying rack for this purpose.

couching cloths

These are the sheets of fabric onto which each newly formed sheet is rolled. Originally, papermakers used heavy felting between each sheet, which is where the name "felts" took hold. Since these are difficult to acquire for most home papermakers, the common mesh kitchen cloths will work well, as will squares of cotton sheeting. Place two mesh towel layers on top of each new sheet before couching the next sheet. Build up a "couching pad" with toweling or pieces of wool blanket that have been washed and put in a dryer to shrink and thicken.

whisk

The fibers of pulp in the vat have a tendency to stick together in clumps, causing uneven sheets of paper to form as the pulp is collected on the surface of the screen. A sturdy metal whisk is used to distribute the fibers evenly in the vat of water. Remove all traces of pulp before they have chance to dry on the whisk.

vat

The vat is the basin or container in which the papermaker combines small amounts of pulp with larger amounts of water. The vat size is dependent upon the size of the work area and the dimensions of the mold and deckle that are being used. There should be a clearance of at least 4in at the sides and a depth equal to, or greater than, the width of the mold. For home papermakers, a dishpan with 14 x 11 x 6in dimensions and a 5 x 7in mold and deckle make a convenient set. A larger vat, such as a laundry sink (22 x 21 x 12.5in), will accommodate a 14 x 20in mold and deckle, but will require a larger working space and drying system. A plastic bathtub raised up on concrete bricks can be used for forming large sheets, while a wooden box lined with plastic sheeting makes a good papermaking vat for any size.

sponge or blotting cloth

This is a useful item to have ready for absorbing excess water during the process of couching a sheet. Water tends to collect in the frame of the mold and impedes the release of the newly formed sheet. Blotting may help to release the fibers from the mesh screen.

materials

pulp

For papermakers, this general term describes any type of prepared plant fiber or recycled paper that is combined with water and used to form sheets of paper.

sizing (or size)

A starch or gelatin solution that is added to the pulp to make the finished paper less absorbent and therefore a better writing or painting surface. It is the final stage in the process of pulp preparation, and occurs after pigments have been absorbed. Liquid size is available and is easily mixed into, and accepted by, the pulp fibers.

retention agent

With the addition of water-dispersed pigments to pulp, a difficulty occurs—both the pulp and the pigment have the same type of electrical charge and so they repel each other. Since the purpose of the pigment is to be absorbed into the fibers, a substance called retention aid or agent is added. Retention agent has the opposite charge and so it facilitates the coming together of the pulp fibers and the pigment.

soda ash

Soda ash is available in powdered form from papermaking suppliers. It is a white powder which, when dissolved in water and used to boil plant material, will help to break down the plant material to allow separation of plant fibre used in papermaking from the rest of the plant.

liquid pigment

Water-dispersed pigments for papermaking are available through papermaking supply centers and are the best for colorfastness, since they unite with the pulp fibers when aided by retention agent (see above).

Retention agent

Sizing

Wood ash

Pulp

Soda ash

Colors and pigments.

sheet is formed, the squeezing out of the water from the new sheets as they are pressed (they are 96% water when they are put onto the press), and the need for safe wiring for the fan in the dryer system. Plan the space and arrange your papermaking area to suit your needs and your available space.

The bending and lifting motions of papermaking can be hard on your back, so try to position work areas to appropriate heights. Avoid slippery floor areas (keep a mop handy!) and wear comfortable clothing and waterproof footwear. When working with pigments and chemicals, wear protective rubber gloves and wash all utensils carefully when you have finished. Strain all waste water before pouring it down the drain. Paper fibers can be hard on plumbing!

If you are spending a lot of time pulling sheets or hanging them to dry, take frequent breaks and stretch out your neck, arm, and back muscles. Standing on a rubber mat while you work will be more comfortable, and much easier on your back!

A wide variety of colors is available. As these pigments may be toxic, care must be taken when working with them. Always wear protective rubber gloves and work in a well-ventilated place.

setup

In order to begin making paper, some form of setup is necessary. Three distinct areas must be established: the papermaking vat, the press, and the dryer system. Arrange the space to accommodate the filling of the vat with water, the dripping of the mold and deckle as each

In this section, a description of some of the many different pulp types available and the specific results produced by each one, will help you decide which type of pulp to choose to begin with. You will learn how to prepare pulp from semi-processed, commercially available products, how to prepare raw plant pulp from plants around you at home and in the garden, and how to prepare pulp from papers that are recycled from your home.

2

PAPERMAKING
techniques

Making beautifully colored paper is a challenge to all paper artists and home papermakers alike. This section teaches you how to pigment your paper using commercially available papermaking pigments, how to use color-loaded papers around you to add color to your pulp, and suggests two ways of making and using your own natural dyes from plants.

pulp preparation

The introduction of wood pulp for papermaking and the development of efficient machines for the production of thin, inexpensive paper revolutionized the world of print and communication. Today, however, the search is on once more for new and alternative materials for papermaking. Old sources of plant fiber are being rediscovered, as the realities of forest management, energy conservation, and the hazards of chlorine bleaching confront our world.

to the instructions in Preparing Plant Fiber Pulp (pages 25–26).

blended, to rehydrate and to break the fibers down into suitable sizes for papermaking.

pulp for papermaking

One of the delightful aspects of hand papermaking is the availability of many different materials from which to make pulp. Since your choice of pulp material will determine the outcome of the finished paper, it is important to know the results that are achieved by using each pulp type.

Plant materials grow all around us and are a good source of fiber for the making of pulp. Experiment with local vegetation—grasses, leaves, some bark and inner bark layers —all can be processed according

cotton pulp

This may be purchased in semi-processed form from most suppliers of papermaking equipment. It comes in large sheets, which must be cut or torn into small pieces and

abaca

A product of the banana plant, abaca is another convenient pulp type, which produces a strong, creamy paper with a fluffy deckle edge, owing to the long fibers of the pulp.

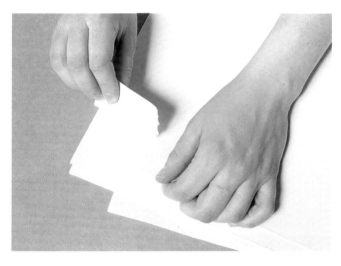

flax fiber

This is beaten in a Hollander beater in order to produce good results. Its high-shrinkage characteristic allows for many interesting effects, including a stiff, rattley paper sheet and wonderful sculptural shapes

hemp

In the ancient world, hemp was used in making rope, baskets, netting, and textiles. It is easier to work with than flax and does not shrink as much. In papermaking, this fiber, when well beaten and processed, will form sheets of paper that are whiter than abaca but not as consistently pure as 100% cotton. It is available in semi-processed form through papermaking suppliers.

thai kozo

This is fiber produced from the dried inner bark of the paper mulberry tree. Its long fiber length and the beautiful tan color make it a desirable plant fiber for papermaking.

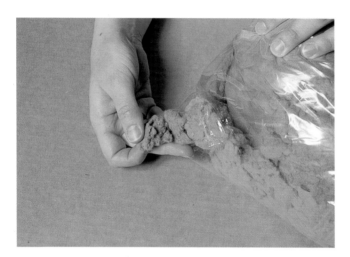

foxfiber

A scoopful of this naturally colored, organically grown raw cotton fiber added to prepared pulp gives the paper an interesting color and texture. Use of a Hollander beater is recommended when using 100% of this fiber, which comes in either a brown or green shade and is available at some papermaking supply companies under the name Foxfiber.

dried leaves

Many varieties of leaves can be found in gardens and forests, both of which are wonderful sources of fiber for papermaking. Leaves must be processed according to instructions on page 26 in order to break down the plant fibers so that they can be beaten into pulp. Fall leaves (with large stems removed) can be made into pulp, which will produce beautiful, fragrant sheets of paper.

recycled paper

For many papermakers, recycled paper is the first source of pulp used in the process of sheet formation. Readily available and cost effective, many kinds of paper can be recycled into a good product for temporary purposes such as envelopes and gift tags. Because of the chemicals inherent in most commercial paper pulp, sheets made from recycled pulp will not be colorfast or lightproof, and so should not be used where permanence is of importance. Newspapers form a gray sludge in the vat and produce dull, muddy sheets.

preparing semi-processed pulp

Plant material that has been partially broken down and formed into flat, dry sheets is called semi-processed pulp. This kind of pulp is available in cotton, abaca, hemp, denim, flax, and others, depending upon the supplier. Most papermaking suppliers carry a variety of semi-processed pulp which is easily packaged and shipped in small or large amounts. The papermaker then soaks the pulp which has been torn or cut into small pieces, and beats it in a blender or other beater system. Semi-processed pulp can be combined with other pulp such as leaf and grass, for a durable and beautiful paper product.

Tear larger pieces of semi-processed pulp into small pieces, about 1in square.

Soak the pulp in clean water for a minimum of half an hour to make it easier on your blender.

Fill the blender ½ to ⅔ full with water and add four or five squares of pre-soaked cotton. Be careful not to overload the machine or it will burn out.

Blend on high power in 15-second bursts. When particles appear to be evenly distributed, with no clumps of fibers, test the mixture using the "Jar Test" (see page 24).

Strain the contents of the blender, storing the prepared pulp in a sealed container. Discard the water, straining it through screening to protect your plumbing. Store strained cotton in a sealed container, or use it immediately to form sheets or to do papercasting (see pages 70–73).

jar test

Pinch off a small amount of processed pulp (about the size of a walnut) and put it into a glass jar filled with water.

1

Close the lid tightly and continuously shake the jar for one minute.

2

preparing plant fiber pulp

Pulp from available plant materials can be processed in two distinct methods. One involves simple cooking to break down the plant fibers and make them more suitable for sheet formation. The other involves cooking the plant material in a caustic solution in order to break down tough fibers and to separate the cellulose part of the plant from the other plant material. Since cellulose absorbs water easily, a high cellulose content is desirable in the formation of a strong sheet of paper.

preparing a caustic solution

Half-fill a pot with fireplace ashes. Cover the ashes with water.

1

Stir carefully over a high heat until the mixture comes to a full boil. Do not breathe in the fumes and always wear rubber gloves for protection.

2

Strain the contents of the pot, discard the undissolved ashes, and put the liquid solution to one side.

3

tip

Soak the plant fibers in water for at least 24 hours before cooking.

preparing the plant material

Gather leaves as they are shed from the trees in the Fall. They can be used immediately for pulp, or they can be dried and stored for later use. Alternatively, process the leaves and store them in measured quantities in the freezer. Large stems must be stripped from the leaves before processing them into pulp. This is a painstaking task but will make the papermaking experience much more successful.

Combine the sorted leaves with the prepared caustic solution and allow to cook on low heat for two hours. **1**

Drain and rinse with clear water two or three times, until all the solution has been rinsed off. **2**

The leaves should tear easily, with no trace of the caustic solution or slimy coating on the surface of the leaves. **3**

In a blender combine 1 cup each of pulp and plant material with water. **4**

Blend in 10-second bursts until the components are evenly distributed. **5**

pulling the sheet

Fill a basin ⅔ full with water. Add 2 cups of prepared pulp.

1

Whisk vigorously to distribute the fibers evenly throughout the water.

2

Using a 5 x 7in mold and deckle, pull a sheet and let it drain for 30 seconds before removing the deckle.

3

Roll the newly pulled sheet onto a prepared couching pad (see page 39), using a smooth motion.

4

If the sheet sticks, blotting with a cloth or sponge will remove excess water and aid in the releasing process.

5

Check the couched sheet for holes before covering with couching cloths or beginning to pull another sheet.

6

The paper is a fragrant blend of cotton and leaves.

For a beautiful, but fragile, variation use 100% leaf pulp.

making grass paper

Ordinary grass clippings take on elegance when processed into pulp and formed into delicate sheets of paper. Choose clippings from a lawn free from chemicals and dogs! The clippings can be used immediately, frozen in 1lb packages for later use, or allowed to break down (a process known as "retting") in a plastic yard bag. Rinse to remove any mold and use as described in this section.

preparing the pulp

Add 4 cups of grass clippings to a stainless-steel pot and pour in enough water to cover the grass.
1

Stir until the water comes to a boil, then turn down the heat and allow to simmer for about an hour, or until the pieces of grass pull apart easily.
2

Strain the contents, discarding the water and rinsing the grass clippings.
3

Prepare 2 cups of cotton pulp according to the instructions on page 23.
4

Blend until the cotton fibers are evenly distributed in the water, then strain the pulp and set it aside for the next step.
5

Combine equal amounts of prepared cotton pulp and cooked grass clippings in a blender with water.
6

pulling the sheet

Place 2 cups of grass/cotton pulp in a basin of water.

Whisk briskly to distribute the fibers throughout the water.

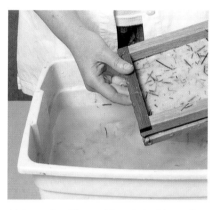

Using a 5 x 7in mold and deckle, pull several sheets of paper, adding pulp as necessary.

The finished sheet will be a stunning blend of cotton and grass.

pulp pigmentation

There are many methods for adding color to paper pulp. Some are as simple as adding a spoonful of powdered paint to the pulp blender, or tearing up gift wrap and adding it to the prepared pulp, encouraging recycling and finding clever ways of using the materials we find around us. As you start to develop your papermaking intuition, it is fun to experiment with a variety of sources. These methods, however, will not produce colorfast or lightfast papers. For a more permanent solution, you will need to use specially formulated papermaking pigments or commercially available fabric dyes.

Whichever method you choose, be sure to follow the guidelines in this section carefully, both for your own safety and to guarantee great papermaking results.

adding color with brightly colored papers

Many common paper products are saturated with color which, when soaked and beaten can be used to form new sheets of paper which are also richly pigmented. Some of the color disappears into the water but much is retained by the pulp. Combine equal amounts of this pulp with beaten cotton pulp for a stronger and longer-lasting product.

Tear the construction paper into small pieces.

1

Soak the pieces for half an hour in order to facilitate blending.

2

Add the soaked pieces to a blender that is ²/₃ full of water.

3

Blend in 15-second bursts until the mixture is evenly blended.

4

Strain the water out of the pulp, which is now ready for use in papermaking or papercasting.

5

Tear colored party table napkins into 1-in pieces. It is not necessary to soak these, as they disintegrate easily in the blender.

6

Choose two or more complementary colors and blend them together in a cup of prepared cotton pulp to give it a mottled appearance

7

Tear pieces of brightly colored gift wrap into small pieces and blend them into a cup of cotton pulp. Add specially chosen metallic pieces at the final stage, blending for a few seconds only, just enough to mix these pieces in, but not so much that the color is lost in the mixture.

8

using papermaking pigments

Papermaking pigments are available from papermaking suppliers and are specially designed to bond with the plant fibers used in the production of pulp. A tiny amount goes a long way so don't be put off by the price. The tip of a small spoon dipped into the pigment is sufficient to color several blender-fulls of pulp, depending upon the intensity of color desired.

Measure the papermaking pigment carefully—a small amount goes a very long way! Here we are using violet- a small amount goes a very long way.

Dissolve the pigment in water, stirring to make sure all the particles are thoroughly mixed in. Do not add pigment directly to pulp, always dissolve thoroughly in water first.

Gradually add the diluted pigment to a blender of prepared, undrained cotton pulp.

Blend in 15-second bursts until the pigment has been dispersed evenly throughout the pulp and each fiber has been coated.

using commercial fabric dyes

Commercial fabric dyes are easily available and work well on untreated cotton, which makes them great for using with cotton pulp that has been processed but not sized. Once the dyeing process is complete, add liquid sizing to ensure color-fastness. Follow the instructions on the package for results.

Measure the powdered fabric dye into water according to the directions on the package.

Sprinkle onto the surface of the water.

Stir until the dye is dissolved completely.

Adding retention agent encourages the attraction between fibers and dye particles.

using natural dyes

Many natural dyes can be made from plants that are readily available in most gardens and from plant products such as tea and coffee. Natural dyes used to color fabric will be equally effective on cotton pulp as well, so it is possible to make matching paper/fabric combinations. Use a common dye bath to make handmade book pages, and a sheet of fabric to make the book cover. Follow the instructions for using a mordant to seal in the natural pigment and prevent color bleaching. Use the same dye bath (here we have used wild cherry dye and a dye made from seaweed) to tint the cotton fabric and the cotton paper.

Experiment with marigolds, beets, and other available plant fibers, and record your results in a papermaking journal.

the method

Before using the natural-dye bath to tint the fabric or the paper, you need to make some initial preparations. The fabric must be treated with a mordant solution, which helps to fix the particles of color to the fabric. (This is similar to the preparation for paper marbling described in Paper Marbling, pages 60–64.) In this instance a solution was made using cream of tartar in boiling water—1 tablespoon of cream of tartar per cup of water.

The fabric was soaked in this solution for 24 hours before being immersed in the dye bath.

To prepare the paper to receive the dye particles, add retention agent to the pulp during beating (see Chapter 1 – Getting Started, pages 13–18).

Meanwhile, prepare the natural dye by boiling 3 cups of plant material (in this example, the seaweed or the wild cherries) in enough water to cover. Remove from heat after two hours and strain, discarding the contents. Add the fabric to the dye bath and allow it to soak for 24 hours, or until the desired intensity is reached. For paper, follow these instructions:

Break up a handful of prepared cotton pulp and add it to the bath of wild cherry dye.

1

Blend with a whisk to produce an even distribution. Soak overnight.

2

Follow the same procedure with the seaweed dye bath.

3

sheet formation

Blend the cotton mixture in its own liquid dye, including the seeds, if desired. Blend using 15-second bursts until well-combined. Do the same with the seaweed pulp.

1

Prepare the basin of water and add 2 cups of the wild cherry pulp. Use a 5 x 7in mold and deckle to pull a sheet. Roll the sheet onto a clean couching pad.

2

Blot with an absorbent cloth to remove excess water and to help the sheet release.

3

Clean out the basin and refill it with clean water and 2 cups of seaweed pulp. Use a clean 5 x 7in deckle to pull a new sheet.

4

Roll the seaweed sheet onto the couching pad.

5

Scoop any excess pulp away from the edges and check for holes.

6

7

Compare the pressed, dried sheet with a piece of fabric dyed in the same dye bath.

couching

Once the sheet of pulp has been scooped from the vat, it must be transferred to the prepared pad without tearing or folding. This step is called "couching" (pronounced "cooching") and requires some practice in order to turn out consistently smooth, intact sheets. Even experienced papermakers have to return sheets to the vat sometimes, where they are whisked into the swirling pulp and reused in the next dip. Carefully prepare a couching pad to increase your chances of success!

tip

Make sure that the surface is smooth—any wrinkles will transfer to your sheet of paper.

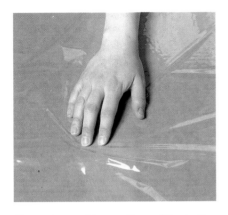

Cover the working surface with plastic, or work on a waterproof table or counter top.

1

A shallow pan, or a lid from a plastic bin, makes a very good couching container, as it will hold the water in places where there is no drainage.

2

Coat a flat board with liquid plastic for waterproofing. This board will form the base of the couching stack.

3

Next, place a thick layer of folded newspaper, toweling, or other absorbent material on top of the prepared board.

4

Cut old woolen blankets into squares to fit the size of the base board, and place two of these on top of the centered paper from step 4. This creates a mound, which facilitates the rolling-off of the sheet from the mold to the couching pad.

5

Cover the woolen sheets with two sheets of heavy interfacing material (available at sewing supply centers) or mesh kitchen cloths.

6

pulling a sheet

This step in sheet formation is so named for the action of pulling the mold through the vat and collecting a thin layer of pulp on the surface of the mold. It requires some practice in order to produce prefect sheets, most of the time. Even beginners, however, can be successful by following the instructions below. If the sheet becomes spoiled during the process, simply touch the surface of the mold to the surface of the water and the pulp will detach itself from the screening. This process is called "kissing off" and is a helpful way to recycle the pulp from spoiled sheets. The best way to determine how thick the layer of pulp on the mold should be, is to practice dipping more or less deeply when pulling the sheet, record your method, and attach a sample of the finished sheet to the method used. This produces a handy reference guide to future sheet formation and takes away a lot of the guesswork in reproducing a particular thickness in the paper.

1

Add about 2 cups of prepared pulp to a basin ⅔ full of water. You will add more pulp as you pull more sheets.

note

This accidental dripping of water onto the sheet was known as "vatman's tears", since the sheet was considered ruined and the vatman was held responsible.

2

Use a wire whisk to disperse the pulp evenly in the water. Be sure to whisk until all the clumps of pulp have been broken up.

Dampen the surface of the mold. Holding the mold and deckle together, with the deckle on top of the screening side of the mold, insert the mold and deckle at the back of the basin.

3

Pull down and through in a smooth motion. The pulp should coat the mold evenly.

4

Continue the movement, lifting the mold and deckle straight up and out of the water.

5

Rest the mold and deckle on the edge of the basin to drain off some of the water. A light tap on the edge of the basin will help to align the fibers.

6

Remove the deckle, lifting carefully to avoid dripping any water on the newly formed sheet.

7

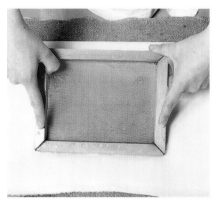

Resting the edge of the deckle on the side of the couching pad closest to you, roll the sheet onto the surface of the couching pad. 8

When the mold is flat against the pad, continue the rolling motion, exerting pressure on the edges of the mold in order to seal the contact between the new sheet and the couching pad. 9

Lift the edge closest to you, and roll the sheet onto the couching pad. 10

tip

Sometimes a bubble will form on an otherwise perfect sheet. You can return the sheet to the vat, whisk, and start again. Alternatively, you can take a pin and puncture the hole, allowing the air to escape, and the bulge to be flattened out in the pressing stage.

pressing

The couched sheet is now 96% water and needs evenly applied pressure in order to remove some of the water and to aid in the binding together of the pulp fibers to form a strong sheet of paper. Pressing times vary according to the thickness of the sheets and the number of pounds of pressure applied. Left too long under pressure, damp paper can become moldy so be sure to remove it when water is no longer dripping from the stack.

Cover the newly formed sheet with two sheets of interfacing. Continue to couch sheets on top of the interfacing, always covering the new sheets with two more sheets of interfacing .

1

The stack of newly formed sheets, called a "post", can be covered with two pieces of woolen blanket before pressing.

2

Cover the post with another coated board (see Couching, step 3), the same size as the base board.

3

Place two bricks or pieces of board underneath the post, to allow air to circulate on all sides of the post.

4

Use C-clamps to secure all sides of the post. Two clamps each side will allow for better distribution of the pressure and therefore more even drying.

5

6

If clamps are unavailable, use a heavy brick placed in the center of the post.

drying

Humidity and temperature affect the drying times no matter which method is used. Air-drying is fast and uses no electricity but the finished sheets may require pressing with a warm iron. Drier systems allow many sheets to be dried at the same time and in a limited amount of space. Paper in a drying system can take approximately 12 hours to dry. To test for dryness, place a hand on the sheet and feel for an even warmth. If there are patches of coolness, the paper requires more drying time.

Hang the pressed sheet, still on the piece of interfacing on which it was couched, on a clothesline to dry overnight. When dry, the sheet can be peeled off and, if a perfectly flat sheet is desired, pressed with a warm iron.

1

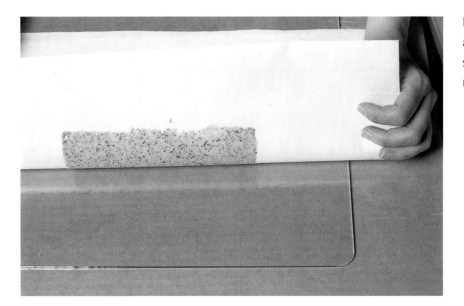

Place the newly pressed sheet, still attached to the interfacing, on a clean sheet of glass (a window will do nicely!).

2

Use a roller to press the sheet onto the glass.

3

Carefully lift a corner of the interfacing and roll it off, leaving the sheet of paper adhered to the glass.

4

If you are very careful, you can lift the newly pressed sheet off the interfacing and roll it directly onto the glass, using a gentle touch so as not to cause tears or wrinkles.

5

The newly pulled sheet can be left to dry overnight on the mold.

6

Use a sharp blade to loosen the edges, and carefully peel it off when dry.

7

To dry a sheet of paper quickly, place the newly pressed sheet between pieces of blanket or toweling and press, continually moving the iron in a rotating motion. This process may cause fading and weakening of the paper fibers, but it is helpful when speed is essential. 8

Form a drying system using layers of corrugated cardboard and semi-processed pulp sheets. 9

Weight the stack down with a heavy brick. 10

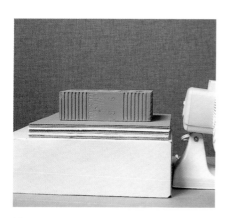

Place an electric fan at one end of the stack, and set it so that the air is blowing away from the stack, and a current is set up, with the damp air being sucked out of the paper stack. 11

Sheet drying on the mold.

One of the delightful features of papermaking is that it can be enjoyed by both inexperienced and accomplished paper artists. The basic preparation requirements and techniques for sheet formation can be learned at all levels. Once you have mastered the basics, however, you will be drawn into the intriguing possibilities in combining color, texture, and pulp types to form unique and exciting papers. Layering contrasting papers, adding

PAPERMAKING
advanced techniques

interesting inclusions, applying spots of colored pulp to a freshly-formed sheet, hiding a flower behind a veil of translucent paper—these and so many more possibilities are yours to explore and create.

Let your imagination go wild and dream up new paper types never before imagined!

laminating

Although it takes a little practice in order to couch one sheet directly on top of another, it is well worth the effort. Laminating is a versatile and imaginative process, since any interesting leaf, cutout, wire shape, or any one of a thousand other possibilities, can be placed on top of the base sheet before the second sheet is layered on top.

For a casual, flamboyant effect, place the second sheet at an angle to the first, creating a unique, eye-catching piece. During the pressing and drying steps, the fibers in the two sheets will bond, and the result will be a single, solid sheet with the laminated object inside, but visible in the light.

Use pigmented pulp to make a base sheet.

Carefully roll the sheet onto a couching pad.

Place a sprig of fern, or a small cedar branch, on top of the base sheet.

1

2

3

Cut two long pieces of contrasting wool or thread and place them ½in in from each side. Press them down to hold them in place.

4

Couch a second sheet on top of the fern and threads.

5

The threads and fern are "sandwiched" between the sheets of paper.

6

Place a flower head on the couching cloth.

7

Couch a sheet of pigmented paper on top of the flower.

8

Lift the mold carefully to show the flower with a thin covering of pulp over it, and allow it to air dry.

9

When dry, remove the finished piece from the couching cloth.

10

Turn it over and use a pin to gently pull the paper open where the flower head peeks through. Be sure to leave enough of a covering to hold the flower head securely in place.

embossing

This process could be called, "Making a Good Impression", since an object is pressed into the paper, or the sheet is pressed on top of the object, in order to create a raised (or indented) impression of the object. The deepest and clearest impressions are formed when a damp, newly couched sheet is used. This sheet should be pressed but not dried.

Keep an eye open for interesting shapes in wire or cardboard, ribbon, netting, children's building blocks, the bubble wrap used in packing, and even the soles of some shoes. Experiment to discover the wonderful textural effects possible in embossing handmade sheets.

Place a piece of textured ribbon on a couching pad.

Couch a sheet of paper over the ribbon at an interesting angle.

Allow the sheet to air dry, under weight, to increase the embossing.

When dry, peel the ribbon off of the finished sheet.

Bend a piece of wire into a flower shape and place it on the couching cloth.

5

Couch a sheet of petal paper on top, aligning the wire shape with the corner of the sheet.

6

Press and dry. When the sheet is dry, peel the wire shape from the sheet, leaving an embossing of the flower design.

7

Cut a heart shape from corrugated cardboard and place it on the couching pad.

8

Couch a sheet of paper on top of the heart. Press and dry.

9

When dry, remove the heart shape, leaving a clearly embossed design.

10

Place a sheet of bubble wrap on the couching pad. Use a heavily pigmented pulp to pull a sheet.

11

12

Couch the newly formed sheet onto the bubble wrap. Allow to air dry, or dry under weights.

layering

Create an intriguing 3D pulp piece for your wall using this technique of couching layers of pulp while removing selected pieces in the process. Combine a variety of pulp types and colors to build scenes and to conceal surprise glimpses into the sheet. Work freely using your imagination. Scoop away little bits of pulp to allow the layer underneath to shine through. You will find it hard to stop creating these pieces, and you can fill a wall with a multitude of pieces, no two the same!

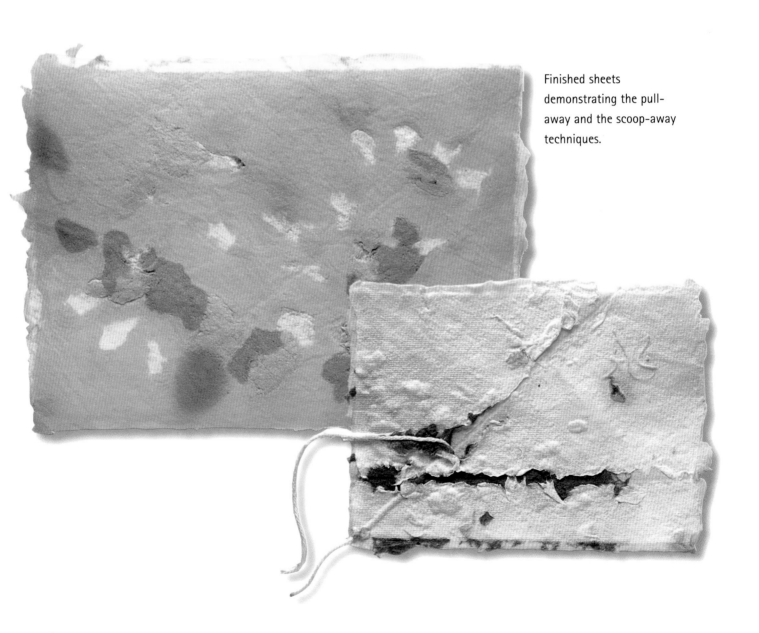

Finished sheets demonstrating the pull-away and the scoop-away techniques.

scooping away

Place small dots of colored pulp on a damp couching pad.

1

Couch a thin sheet of yellow pigmented pulp on top of the dots of pulp.

2

The colored dots will be visible through the newly formed sheet.

3

Scoop away some sections of the yellow top sheet.

4

Couch a second sheet of contrasting pulp (here we have used cotton and petal) on top of the yellow sheet.

5

pull-away

On a newly formed yellow sheet, couch a triangle of blue pulp.

tip

Simply scoop the blue pulp onto the mold without the deckle and place it where desired.

Place a 12in thread 1in in from the top and bottom edges. Couch a second sheet of a contrasting pulp (cotton and petal).

tip

For a finer line and easier control of the pulling process, the piece may be dried before the threads are pulled.

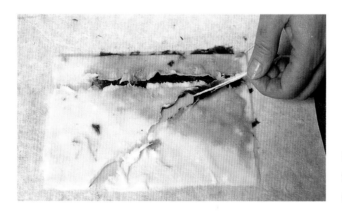

When the two sheets have been pressed together, gently pull the threads through the damp sheet for a stunning textural effect.

crumple paper

What do many of us do to release tension or to control a burst of emotion? We wad up a piece of paper and toss it at the wall! Here is a healthy and delightful use for the results. Actually, the process of paper crumpling has a noble history in Japan, where the art of "Momogami" ("crumple paper") is an intricate and beautiful technique. Here is a much-simplified version of the process, which produces a sheet of paper with structural effects of another dimension. When combined with a fold-and-dye technique, the texture is emphasized. Try using a marbled sheet for a beautiful finished product. Because the crumpling is done with a damp sheet, the lines will be permanent and these can be used for interesting sculptural effects. The sheet can be mounted for display, or used as a wrapping around a handmade book or fragile gift. As you see the unique results achieved by paper crumpling, you will think of many new ways to use these sheets.

Couch a fresh sheet of handmade paper. Here, we have made a sheet of black cotton.

1

Press the sheet to remove excess water and encourage the fibers to bond together.

2

When pressing is complete, carefully peel the sheet from the couching fabric. The sheet should be damp and easy to handle.

3

Crumple the sheet into a ball. The more crumples, the finer the detail will be at the end of the project, so don't be afraid to scrunch the paper into a tiny ball.

4

Place the crumpled paper into a plastic bag.

5

Tightly seal the bag with an elastic band or string. Put to one side for 24 to 48 hours.

6

After the appropriate time has elapsed, open the bag and remove the paper.

7

8

Flatten the paper by hand and allow to air dry. The paper will be crinkly with texture.

paper marbling

Marbling is another method of applying color to pre-formed sheets of paper. Like so many papermaking and decorative-arts methods, marbling has been said to have its roots in 16th-century Persia, but the technique has now re-emerged, as contemporary papermakers rediscover the fascinating effects of marbling on handmade paper. "Suminagashi" is the name given to the Japanese technique of paper decorating, which may have been in practice as early as the 12th century. The word suminagashi, or "ink-floating", refers to the practice of carefully dropping colored inks onto a water surface. The colors were then blown across the surface to create beautiful, delicate swirling patterns. When a sheet of paper was placed on top of these patterns, a print was made. Commercially available marbled papers in a variety of colors and arrays give some idea of the range and intricacy possible in paper marbling. Some preparation is required, but the basic method produces excellent results, even for total beginners.

The marbling technique described here is a mono-printing technique, in which colors are deposited onto a gelatin-based size. By using sharp tools, rakes, hair picks, combs, and a specially-designed stylus, a pattern is formed on the surface of the size. Paper is treated with a mordant to fix the colors, and the paper is then placed on the surface of the size. When the paper is removed, the pattern will have been transferred to the paper. The surface of the size is then skimmed with a strip of newspaper to remove any remaining colors, and the process is repeated. No two marbled sheets are alike!

Before beginning you will need to assemble some basic equipment:

The **bath** or tray should be 3 to 4in deep and at least 1½in larger than the dimensions of the paper to be marbled. Metal or foil roasting pans, kitty-litter trays, or a waterproofed wooden vat all make good containers. It is helpful to have a light-colored vat in order to see clearly the colors floating on the surface.

The **size** or sizing agent provides a gelatinous surface on which the colors are deposited. There are several options, but here we have used carrageenan powder, a seaweed derivative available from marbling suppliers. Add 1 tablespoon of powder to 4 cups of warm water and mix in a blender. Empty this into the vat and repeat the process until a depth of about 2in has been reached. For best effectiveness, allow this to stand overnight. Experience will tell you if the size is too thick or too thin. To begin with, remember that colors will not spread if the size is too thick: they may even sink to the bottom of the vat. If the size is too thin, the colors will disperse too much, you will not be able to form patterns, and the hues will lose their brightness. Practice and record your results to help you achieve the effect you want.

The **colors** used here are acrylic paints, which are available in many shades; effective in paper marbling, and readily available from artist supply stores. Thin the paint with distilled water until it reaches the consistency of single cream.

Ox gall comes from the gall bladder of a bovine, and is combined with the colors because it breaks the surface tension of the size and allows the color to spread. Using an eyedropper, add to the color, one drop at a time, experimenting to see how many drops are required.

The **mordant** is used to treat the paper before marbling, in order to fix the colors to the surface of the paper and to improve colorfastness. Alum is used here. It is readily available from any drug store or grocery store, and it forms a slimy surface that adheres to the paper when it is lifted from the bath. Add 1 tablespoon of powdered alum to 1 cup of boiling water. Stir to dissolve completely and allow to cool.

Marbling tools can be easily fashioned from readily available objects. Collect a few of these before you begin and plan how you will use them. A stiff whisk can be made by tying together several long bristles from a straw broom. A marbling comb can be made by taping straight pins onto a strip of cardboard and placing another piece of cardboard to cover the heads of the pins. Duct tape will help to preserve the comb when it gets wet.

tip

To test the amount of ox gall you have added to the color, drop a few drops of color onto the surface of water in a small bowl. The color should spread across the surface of the water and form interesting swirls and patterns. If the color just remains where it falls, add more ox gall to the color mixture.

Marbling tools
- combs: including hair pick, wide and narrow combs
- acrylic marbling pigments
- grass whisks
- plastic tray
- strips of newspaper
- ox gall in bottle with dropper top

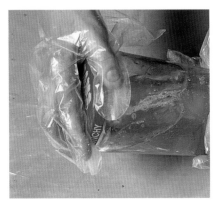

Prepare the mordant by mixing 1½ oz of alum with 1 pint of boiling water. Stir well to dissolve the crystals. Place in a glass jar and cover with lid. Allow to cool. Shake jar well before using.

1

Cover your work area with plastic. Using a sponge, and working up and down the page, apply a thin layer of mordant to the paper—just enough to dampen the paper.

2

Now apply the mordant across the page. When finished, mark the other side to remember which side has the mordant. Place the paper between two boards, with a weight on top to prevent the paper from cockling (warping) during drying.

3

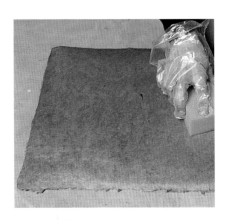

Repeat steps 2 and 3 with other sheets, to have several ready for marbling.

4

Using a clean jar, prepare the colors following the instructions on pages 60–61. Use a whisk to blend well.

5

Prepare the size according to the directions on page 60. Pour into the marbling tray. "Charge", or prepare, the vat to accept the pigments by skimming the surface of the vat with a strip of newspaper. Shake small droplets of pigment onto the surface of the size by tapping the whisk against your hand.

6

Apply another color in the same manner. Add as many colors as desired—generally, two or three colors are a good start.

7

Draw the comb toward you, across the surface of the size, creating a swirling pattern. Be sure to keep the comb at a slight angle, and keep the comb tines just deep enough to make contact with the size. Move slowly and evenly, keeping your arms and hands as comfortable as possible.

8

Repeat, drawing the comb across the surface to create a contrasting swirl.

9

Retrieve a pressed and dried sheet of mordanted paper and place one corner of the paper (mordant side down) onto the surface of the water. Slowly but smoothly ease the rest of the sheet down.

10

Carefully peel back the sheet from the vat.

11

Place the sheet on a board.

12

Rinse well with fresh water to remove the size.

13

Hang the sheet to dry.

14

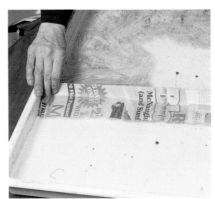

Clean the vat by skimming the surface of the size with a strip of newspaper. Repeat several times to remove all pigment.

15

Create freehand patterns by using a stylus to add swirls and loops where desired.

16

A hair pick is also a useful tool for making patterns in the vat.

17

A bubble pattern is created by applying the pigments evenly and leaving them uncombed.

18

Place the paper onto the surface of the vat and then carefully remove it.

19

Rinse well. The white cotton paper makes a lovely base for this color scheme.

20

tip

If the pigments wash away during rinsing, you have marbled on the non-mordanted side of the paper.

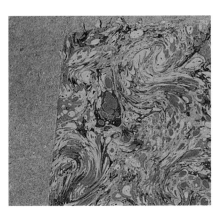

Experiment with pigmented sheets of paper. This sheet of sage green abaca paper looks striking with the yellow and red pigments.

Assorted marbled papers.

using pigments and inclusions

The beauty of choosing and mixing your own pulp for papermaking is that there are so many wonderful options! Beyond pulp type and color, there is the possibility of adding pigment and inclusions to the pulp. By inclusions, we mean adding any small bits or accents that will enhance the paper, or accomplish a desired effect.

Natural inclusions such as flower petals, dried or fresh herbs, or even coffee grounds may be used to add texture and interest to your papers. Lavender adds long-lasting scent and is ideal for creating personal stationery products, unique to you!

pigmented paper

abaca pigmented with fabric dyes

These beautiful papers were colored with fabric dyes. When using fabric dyes, dilute the powdered dye in water and add to the pulp according to the following directions. Experiment with your own combinations of dyes and dye strengths to create your own unique papers. Look for fabric dyes in the housewares and craft departments of local stores. Sizing should be added once the paper has been fully pigment, just before pulling sheets.

orange abaca paper

To $1/4$lb beaten abaca pulp, add 2 cups of orange fabric dye mixed to full strength according to manufacturer's directions. Blend the pulp and dye together and allow to sit for several hours. Add retention agent if needed.

pink abaca paper

To $1/4$lb beaten abaca pulp, add 2 cups of red fabric dye, mixed to half strength according to manufacturer's directions. Blend the pulp and dye together, and allow to sit for several hours. Add retention agent if needed. A darker pink color can be created by mixing red fabric dye to $2/3$ or $3/4$ strength.

purple abaca paper

Follow the instructions for orange abaca, adding 2 cups of full-strength purple dye to $1/4$lb beaten abaca pulp.

fuchsia abaca paper

Follow the instructions for orange abaca, adding 2 cups of full-strength fuchsia dye to $1/4$lb beaten abaca pulp.

papers colored with construction paper

A fun project for beginners and advanced papermakers alike, is paper colored with torn bits of construction paper. Because construction paper is highly pigmented, it breaks down to add overall color to the paper, as well as flecks of a particular color. By adjusting the blending times, you can control the size of the fleck. Long blending produces a smooth, evenly colored pulp, while shorter bursts will result in chunky pieces.

red-flecked paper
Soak ⅛lb recycled pulp for 10–15 minutes. Place the pulp in a blender and add two sheets of red and one sheet of blue construction paper, torn into 1 x 1in pieces. Lightly blend for five seconds. Check that all the construction paper has been blended, with large flecks visible.

yellow-and blue-flecked paper
Follow the directions for red-flecked paper, but add two sheets of yellow and one sheet of blue construction paper torn into 1 x 1in pieces. After blending, the paper will be a mottled green with visible flecks of yellow and blue. Blend for longer if a smoother green pulp is desired.

dark gray paper
To ⅛lb soaked recycled paper, blend six sheets of black construction paper, torn into small squares. Blend the pulp in several short bursts of five seconds, until the paper is dark gray with subtle flecks.

blue paper
Follow the directions for dark gray paper, adding four sheets of blue construction paper instead of black, torn in 1 x 1in squares.

botanical inclusions

Adding botanical inclusions to paper is a lovely way to capture fresh and dried herbs and foliage, letting you enjoy the beauty of nature year-round. Remember that green stems will bleed green or brown in your paper, and that white petals often turn brown. Experiment with different herbs and flowers to find your favorites.

lavender buds in cotton

Blend ⅛lb cotton pulp and add 2 tablespoons of lavender buds, mixing the buds in by hand. Transfer the pulp to a vat and blend pulp well with a wire whisk. Lavender buds may bleed a slight green or yellow, adding extra interest to the papers.

light purple abaca with lavender leaves

Blend ⅛lb light purple abaca, using either fabric dyes or papermaking pigments. Hand-blend in approximately 2 tablespoons of fresh lavender leaves. Pull sheets of paper and press to dry both paper and leaves.

fresh moss paper

In ⅛lb blended recycled paper pulp, add 1½ cup of fresh moss cut into ½in pieces. Pulp lightly for two to five seconds in a blender to incorporate the moss. Whisk further once the pulp is in the vat.

spanish moss and petal paper

In ⅛lb blended cotton pulp, hand-mix ½ cup Spanish moss cut into pieces of several inches in size, along with 4 tablespoons of fresh or dried petals. Add to the vat and pull sheets. Press and dry as usual.

pink rose petal paper

Blend ⅛ lb recycled paper pulp. Spray ¼ cup of dried pink rose petals with spray fixative. Alternatively, use fresh petals and do not spray. Add petals and recycled paper to blender, and pulse for one second. Avoid overblending, as this will

pulverize the petals into tiny bits.

fresh parsley

Large leaves of fresh parsley are striking inclusions in hand papermaking.

Add ¹/₄ cup of fresh leaves to ¹/₈lb cotton pulp and blend by hand until thoroughly distributed. This recipe will make sheets of paper speckled with beautiful parsley leaves. Try with other fresh herbs such as basil and thyme.

yellow statice paper

Make this sunny yellow paper by harvesting purple statice, and picking the buds from the stem.

Add ¹/₂ cup of purple statice to ¹/₈lb yellow cotton pulp, blending for one second to incorporate pulp and petals.

fern and petal paper

Prepare ¹/₈lb abaca pulp and add the following: 2 tablespoons each of yellow, red, and purple petals, ¹/₂ cup fresh fern cut into 1–2in pieces. Mix in blender for one second to lightly blend. Makes a very colorful and interesting petaled and ferned paper.

papercasting in molds

Papercasting is the technique of producing a paper version of an original shape. The cast is the finished product that has been formed or taken from a mold. Some shapes and patterns seem to call out to the papermaker to reproduce them using the technique of papercasting. Because of differences in fiber characteristics, as well as variations in beating time and methods, some experimenting should be done, and notes kept regarding the results. Cotton pulp is an excellent material for casting, since it will pick up fine design details, and the final product is lightweight and durable.

Use found objects such as candy-making molds, commercial papercasting molds, or bubble-wrap to fill with a layer of pulp. Or, cast the pulp on the outside of the object (but note that this will be more difficult to remove).

Try using modeling compounds, or plaster of paris, to make your own molds using special objects—a clam shell, a large-veined leaf, a medallion, or a coin—any object that can be pressed into the compound and removed, leaving an interesting shape or design.

tip

Experiment with the amount of sizing desired. Generally, the proportions used are 2 tablespoons of liquid sizing per 1/2lb of pulp for a medium.

Prepare the cotton pulp just as you would for papermaking. *1*

Select a terracotta mold and spray lightly with vegetable oil. *2*

Place a scoop of pulp onto the mold and push into all the crevices. *3*

Continue adding pulp until the mold is covered. *4*

With a dry cloth, blot the cotton pulp, pushing it further into the mold as you remove excess water. Blot until water no longer comes out. *5*

Allow the casting to air dry, place in a food dehydrator or microwave for 30 seconds at a time. (It may take several sets). Press the pulp into the mold with a dry cloth after each one. *6*

When the casting is dry, use a blunt knife to remove it from the mold. *7*

tip

If are intending to paint the finished piece, add some sizing to reduce the absorbency and bleed of felt-tip markers and paints.

types of mold

Other terracotta molds have multiple small designs on one mold. Cast all four at a time and then cut them apart, or cast them individually as shown.

Create your own molds by using plasticine and other pliable modeling compounds. Select an object with good 3D relief, such as this shell, and press it firmly into the modeling compound.

1

Remove the object from the compound.

2

Fill the cavity with cotton pulp. Allow to air dry and remove.

3

Remove the pulp from the mold.

4

tip

Place the mold over a bucket when removing water, to catch any drips and to intermittently squeeze out the blotting cloth.

plaster of paris

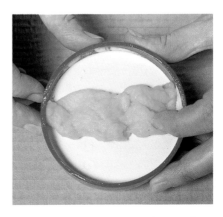

Another excellent mold-making material is plaster of paris. Mix the plaster according to the package instructions and pour it into a plastic container. Press the object that you are casting into the plaster and leave to dry.

1

When the plaster is dry, remove the cast object.

2

Fill with cotton pulp and allow to dry.

3

When dry, remove the casting from the mold.

4

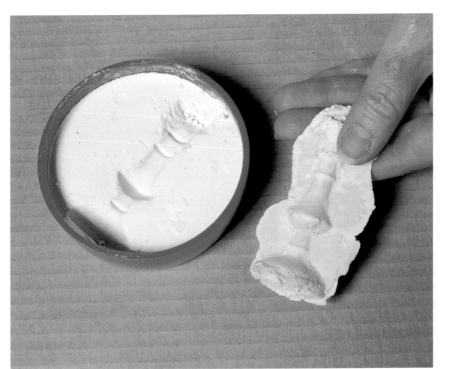

cast paper bowls

Cast paper bowls are not only decorative, but may be used to display your treasures—colored glass, pot-pourri, seashells from a special holiday, or some shiny stones. The possibilities are limitless for both form and function as you indulge your imagination in creating bowls with different shapes, sizes, and inclusions. Use threads, grasses, raffia, or fine wires, leaving the ends to extend beyond the rim of the bowl.

Mix pulp types and colors, add sparkle or flower petals to the pulp, or form a classic, single-toned piece. Incorporate fabric pieces and stitching for unique textural details. Look around for interesting bowl shapes such as light fixtures or plant pots.

Don't limit yourself to the kitchen—try hubcaps or children's sand toys. Most of all, have fun with this project, which is one that is suitable for all ages, and for novices as well as experienced paper artists.

Apply a thin layer of petroleum jelly to the inside of a bowl. Pull and press fresh sheets of handmade paper, or dampen dried sheets. Tear into small pieces, about 2 x 3in, and apply to the bottom of the bowl.

1

Add more sheets, building up the base of the bowl.

2

Continue adding more pieces of paper, until the entire bowl surface is covered.

3

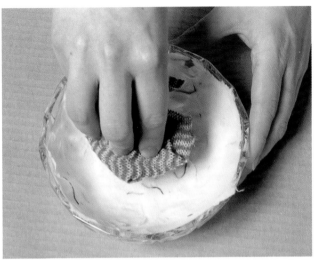

Using a damp cloth, blot the paper to bond the fibers together. Put aside to air dry.

When the cast bowl is dry, remove from the piece by gently loosening the edges with a blunt knife. Start first with one edge, and then ease the knife between the paper and the bowl, until the entire piece is loosened.

tip

If the paper starts to dry out before you have built up all of the layers, mist it lightly with water.

variation

Create a chunkier bowl by using pulp applied to the bottom of a bowl. Apply a thin layer of petroleum jelly to the bowl. Take a small amount of pulp, gently squeezing out any water so that pulp is damp, but not dripping wet. Press into the walls of the bowl. Repeat until the bowl is covered. Press with a damp sponge or cloth to bond the fibers together.

1

Add details such as contrasting confetti. Press into the pulp.

2

3

Add a few small pieces of pulp to hold the confetti in place.

watermarks

The development of watermarked papers first appeared in Italy in the 13th century. During the next two centuries this became a popular marking throughout Europe, and papers with images of animals, fruits, and images from nature and religion—stars, moons, crosses—helped to identify individuals and papermills. Today, watermarks continue to be used to mark personal stationery, and traditional methods are still in practice. In industry, mechanization of the process allows for large quantities of watermarked papers to be manufactured, but you can achieve fine results at home by using the techniques described in this chapter.

The watermark is really a misnomer since water does not create the image in the paper. The watermark is a translucent design in the paper, formed in the making of the sheet of paper by attaching a wire or thread design to the mold surface before couching the sheet. As a result, the pulp is distributed more thinly in the raised area than throughout the rest of the sheet and, especially when held up to a light source, the design appears as a lightness within the dry sheet.

Here you will learn about some different materials used in the making of watermarks. Try forming a wire design sewn onto the screen, "draw" a design directly onto the mold using fast-drying white bond glue or hot glue from a glue gun. Cut shapes from adhesive tape and stick them onto the surface of the mold.

Experiment with these methods until you achieve the best watermark, either to create your personal line of stationery products, or to make a unique gift for someone special.

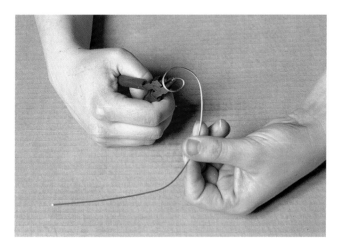

Using needlenose pliers and medium-gauge wire, create a loop.

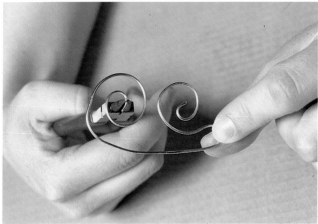

Finish off the piece by looping the other end into a swirling design.

Thread a fine needle. Position the wire piece as desired on the right side of the mold and sew in place.

2

Place the deckle onto the mold and prepare to pull a sheet.

3

Pull a sheet of paper and drain.

4

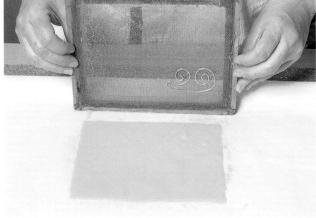

Couch the sheet. The sewn wire piece will leave a subtle impression.

5

A hot glue gun also creates an effective watermark.

6

Pulling and couching a sheet with a watermark made with hot glue yields very similar results to a wire watermark. Depending on the type of screen on your mold, the glue may or may not be removable.

7

Mask off an area of the mold to create a watermark. Here, we applied carpet tape in the shape of an "H".

8

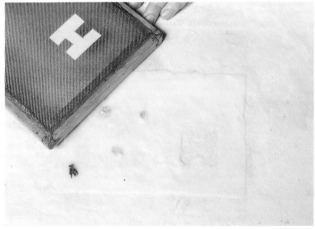

Couch the watermarked sheet and then couch another sheet directly on top. Without the second sheet, the "H" will be a cutout rather than a watermark.

9

pulp painting

Pulp painting is a wonderful technique for creating imaginative and colorful designs and pictures, formed directly on the surface of the mold. Since the pulp fibers of each pulp type bond as they dry, the finished product is a single piece, which, when air dried without pressing, has a roughly textured face and a smooth backing where it has been against the screen. This allows for easy mounting and display of your work of art!

Use a variety of pulp types and colors, making sure that the pulp has been well beaten to make it easier to manipulate. Very fine pulp may be put into squeeze containers and applied to a newly formed base sheet. The method used here included simply spooning pulp onto the mold, arranging the fern fronds, and applying enough pulp to hold them in place. It is fun to "paint" landscapes using pulp, because the pulp itself creates features of depth and variety.

Prepare several colors of cotton pulp. Put into squeeze bottles or bowls.

1

Squirt or place colored pulp onto inverted mold.

2

Add fern fronds to the mold and anchor them in place with pulp.

3

Continue applying pulp to the mold until covered. Fill in any gaps using tweezers or a small stick.

4

Press the pulp well to absorb water and bond the pulp fibers together. Let it dry thoroughly.

5

When dry, carefully remove from the mold.

6

envelopes

When creating handmade writing paper and stationery items, envelopes are a necessity. Using the technique of masking to alter the shape of a rectangular mold allows you to custom size your envelopes to match the sheets you have made. Or, make your own envelope deckle to apply to the surface of the mold to form envelope sheets. Vary the shapes and sizes to suit your needs, and experiment with pulp types and colors. A bone folder is helpful in forming crisply defined edges. Envelope glue may be applied to the flap, and is available from papermaking suppliers. Alternatively, simply seal the envelopes using a glue stick.

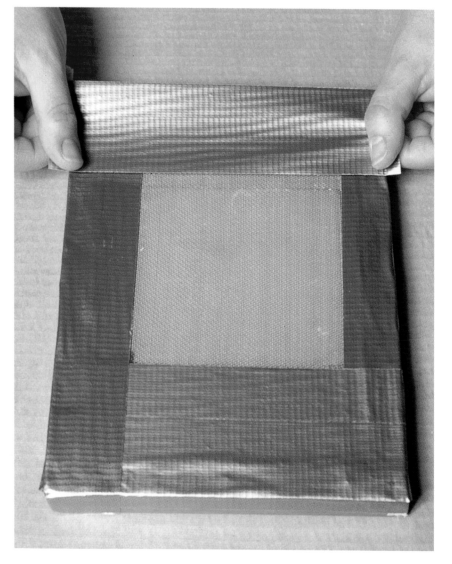

Using duct tape or other waterproof tape, mark and mask a deckle to create the desired shape and size. This technique works well when making many sheets of a particular size.

1

Prepare a template using the diagram shown. Cut out using scissors or a craft knife.

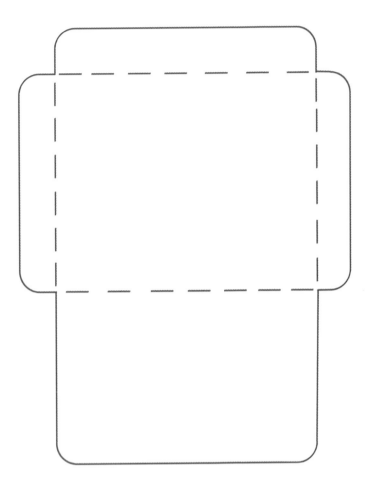

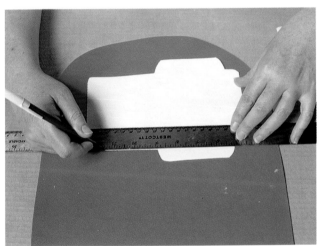

Place the template on a flexible plastic placemat. With a pen or pencil, trace around the template. Remove the template and cut out deckle.

2

Put the envelope deckle between the pieces of an 8½ x 11in mold and deckle.

3

Holding the three layers together, pull a sheet of paper. Remove the deckles and couch sheet.

4

Press and dry.

5

When dry, fold the envelope flaps.

6

Apply a thin layer of adhesive to the side flaps and smooth into place.

7

tip

Label the envelope deckle and save it for future use.

fold and dip paper

Follow these instructions to create a beautiful fold and dip sheet. Once you are familiar with the technique, you can vary the pulp type of the base sheet, work with different color combinations, vary the fold size, and try using more or less water to dampen the sheet. You will discover a kaleidoscope of results, all startling in the patterns that emerge and the display of color! Use these papers to line the inside pages of handmade books, or to cover the outside of the book. Wrap very special gifts, or try stitching around some of the lines to accent the texture of the folds and mount the sheet for display.

The use of a bone folder allows for crisp, permanent folds in the paper and more control over the application of color. There is a fine balance between controlling the process and allowing the medium to develop freely, with reckless abandon. A combination of careful planning and preparation will allow for the best flow of the color to all areas of the sheet.

Dampen a sheet of handmade paper. Fold lengthwise to create a fan with 2in pleats.

1

Use a bone folder to create crisp edges.

2

Starting at one end, fold a triangle. Crease the triangle with a bone folder. Fold the triangle back and continue, back and forth, until the entire strip is folded.

3

Clamp the triangular bundle at one corner.

1

Pour three different colors of ink into a container. Dip the corners of the bundle into the ink, moving the clamp as necessary.

2

tip

Experiment with smaller and larger pleats to create different patterns in the finished piece.

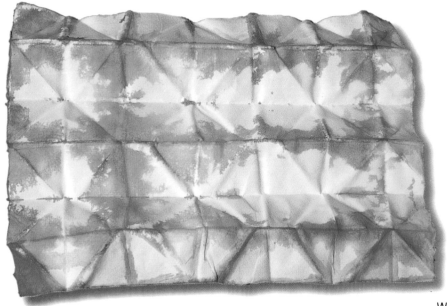

3

With a clamp holding the triangle together, allow the sheet to dry overnight. When dry, remove the clamp and smooth out the sheet.

Now that you have perfected your skills in sheet formation and have mastered some advanced techniques for creating completely original paper types, what can you do with your new talent? Here are some projects to start you off on a whole new journey of papermaking with a purpose. Learn some basic and decorative binding techniques for producing your own handbound books with handmade pages and covers. Craft a

PAPERMAKING *projects*

beautiful photo frame for personal display or for a thoughtful gift idea. You'll never want to purchase another greeting card— you can make your own, unlike anything else on the shelf!

All this and so much more to turn those wonderful sheets into works of art!

paper sampler book

A paper sampler book is a great introductory bookbinding project, and a wonderful way to record papermaking recipes and notes. Keep all your findings in one place. Include samples of the papers you have made, notes about your exact techniques, suggestions about what you'd do differently the next time, and recipes to allow you to reproduce great results every time! This book will become a treasured and well-used addition to your library. Use quality materials to ensure that your book will last for many years.

materials required

five sheets of 8½ x 11in
        handmade paper
one sheet of thicker 8½ x 11in
        handmade paper
colored embroidery thread
bookbinding or embroidery needle
beeswax
10 paper swatches
adhesive
clip

sewing the book

Select five sheets of handmade paper for inner pages and one sheet of contrasting handmade paper for the outer cover. The outer cover should be of thicker weight than the inner pages, to give it strength. Here we are using white cotton inner pages and a shimmering mint cotton cover, all in 8½ x 11in.

1

Place the inner sheets in a stack with the cover paper on top. Using a bone folder, fold the stack in half. Crease the fold well using the bone folder.

2

Using a scrap piece of paper, create a template by marking a center hole and a hole 2½in from both the head and tail of book. Open the book to the center, and clip the template along the fold. Using a pencil, mark the holes on the fold of the book.

Remove the template and re-clip the book to keep the pages together. Pierce the marked holes using a blunt bookbinding or embroidery needle, or a fine awl.

Cut a length of colored embroidery thread several feet long. Wax the thread by pulling it through beeswax. Waxing helps the thread tighten and hold in place during binding. Thread a needle and begin sewing by entering the center hole from the middle of the book.

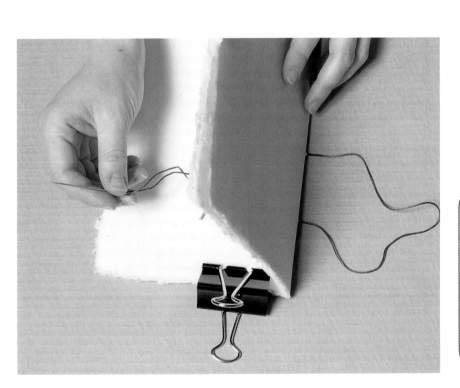

Pull the thread through the center hole, leaving a 3in tail in the center of the book. The tail will be used at the end of the project to tie off the binding. From the outside of the book, enter the top hole. Pull the thread until taut.

tip

The top of the book is called the head and the bottom of the book is called the tail.

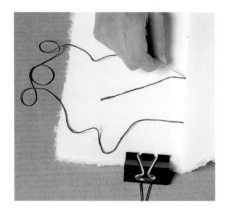

Re-enter the center hole and pull taut.

From the outside of the book, enter the bottom hole.

Tie a double knot over the top of the center hole.

9

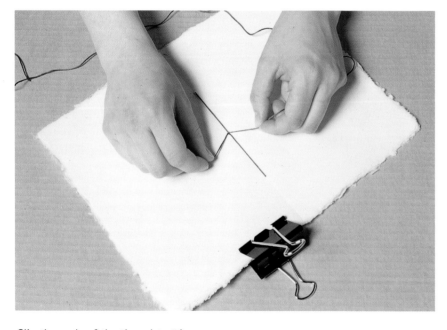

Clip the ends of the thread to ½in.

10

tip

Label the template with a project name and keep for re-use. A file folder, a small box, or clear plastic sleeves are great for organizing bookbinding templates.

adding tipped-in paper swatches

Select nine paper swatches. Here we have selected two groupings of samples—pigmented papers and plant-fiber papers.

Cut or hand-tear swatches into approximately 3 x 3in pieces.

Apply adhesive to the back of the paper. Use either dry roller adhesive, white glue, or photo mounts.

Place the swatch on the second page of the book, positioning it 1½–2in from top, and centered from side to side.

Continue adding swatches to the book in the same manner.

Below the swatch, record the recipe and other notes about the paper swatch.

Trim a piece of contrasting handmade paper to 2 x 3½in. Apply adhesive, and place the piece on the front cover of the book, centered and 2½in from the top.

Trim a second piece of contrasting paper to 1 x 3in and layer on cover.

8

stab-bound books

Stab binding is also referred to as Japanese binding and is made using two covers with single or folded inner pages. Traditional stitching patterns include the hemp-leaf, the tortoiseshell, and many others, each bearing a legendary explanation. For this book, we have chosen a basic five-hole binding technique. It is a flexible and adaptable binding style, which works with all book sizes. Covers can be soft or hard. Hardcovers can be made from heavy card stock or bookbinding board. Use a hammer and awl for punching or "stabbing" the holes, or drill them using a craft drill and a fine bit.

materials required

12 sheets of 5 x 8in
   handmade paper
two sheets of 5 x 8in handmade
   paper in cover weight
two clips
awl
hammer

softcover

Place the inner pages on top of the back cover, lining up the left edge. Place the top cover on the stack.

1

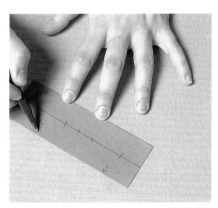

Create a template from a piece of card stock the same length as the binding edge. Here, the template measures 2 x 5in. Draw a line 1in from the edge. Mark the edge with arrows. Mark holes at 1, 1³/₄, 2¹/₂, 3¹/₄ and 4in from the top of the template.

2

Clip the template to the edge of the book. Make sure the edge with the arrows is lined up with the edge of the book.

3

Using a hammer and awl pierce the five marked holes.

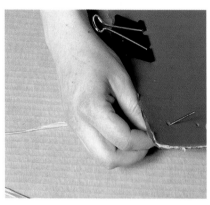

Thread the needle with a length of raffia. Enter hole #1 from the back of the book. Pull the needle and raffia through, leaving 3in of raffia at the back of the book.

5

Loop the raffia around the spine of the book and re-enter hole #1.

Pull the raffia until taut.

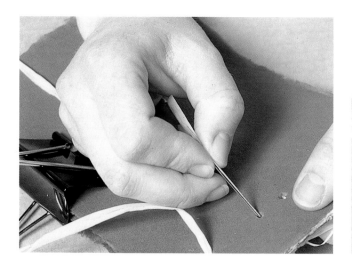

Enter hole #2 and pull the raffia through. Loop the raffia around the spine of the book and re-enter hole #2.

8

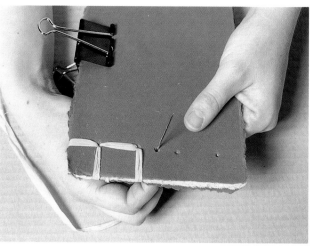

Enter hole #3 from the back of the book and pull through. Loop around the spine and re-enter hole #3.

9

Enter hole #4 from the front of the book, pulling through, looping around the spine and re-entering the same hole.

10

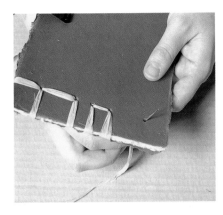

Repeat the last step with hole #5.

11

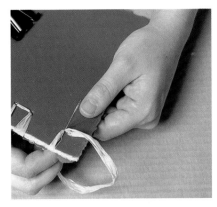

Loop the raffia around the tail of the book and re-enter hole #5.

12

Sew back up the book to fill in the blank areas, starting by first entering hole #4 from the front of the book.

13

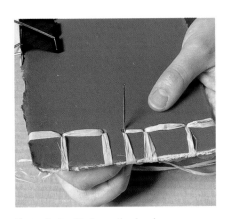

Enter hole #3 from the back.

14

Repeat the last step with hole #2, entering from the front of the book.

15

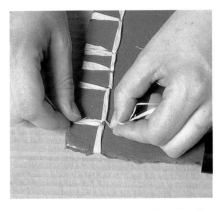

Tie the raffia on the back of the book over hole #1, using the length of raffia left in step 5 (see page 94).

16

hardcover photo album

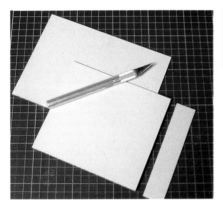

From the bookbinding board, cut three pieces measuring 5 x 1in, 5 x 5½in and 5 x 7in.

materials required

brayer
bookbinding paste
18 sheets of black acid-free paper
Two sheets of handmade paper, 6 x 9in or larger
6 x 9in or larger
craft drill
raffia
bookbinding or embroidery needle

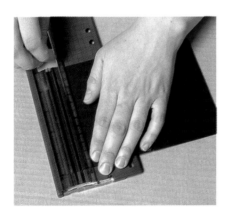

Trim the black paper to 5 x 7in for the inner pages.

2

Select the paper for the cover. If using petal paper, where the petals are more prominent on one side, place that side face-down. Using a pencil, trace the edges of the 5 x 7in board on unpetaled side.

3

Apply paste to the paper.

4

Position the smallest boards within the outline of the back board, so that the 5 x 1in piece is flush with the outer edge, leaving a gutter of ½in.

Flip the sheet with the adhered boards over, and smooth out any wrinkles using the brayer.

6

Apply a thin layer of paste to the corners, and miter by folding in at a 90-degree angle.

7

Apply paste to the 5 x 7in sheet of matching paper, and position on the uncovered back of the board.

8

Use a brayer to smooth in place. Repeat steps 4–9 with the back cover, this time using one piece of board, and not creating a gutter for the hinge.

tip

Place a piece of scrap paper between the clip and the book to protect the book from getting marked.

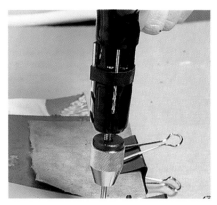

Jog the black inner pages until all the edges are even, and place them between the front and back covers. Clip in place. As in the softcover binding project (see page 93), create a template with two holes positioned ⁵/₈in from the edge and 1¹/₂in from the head and tail.

9

Clip the template to the front of the book. Using a craft drill with a fine bit, drill holes as marked.

10

Thread the needle with ¹/₈in-wide yellow satin ribbon. Start sewing from the front, looping around the spine and re-entering the hole, then looping around the head and entering the same hole for a third time. Leave a 3in tail and pull the ribbon taut. Enter hole #2 and loop around the spine and tail, pulling the ribbon taut. Bring the ribbon up to hole #1 and tie the ends in a knot. Finish with a decorative bow. Trim the ends.

11

tip

Save sheets of handmade paper that have a blemish in one corner for covering books. Since the edges are often trimmed when covering the binding board, the blemish can be positioned to be removed and the sheet put to good use.

photo frame

Handmade paper provides rich and decorative material for framing any photo or piece of artwork. The variety of colors and textures means that you will be able to coordinate the frame with the photo in order to enhance the overall effect of the completed product. Use grass paper to frame a summer scene, for example. Work carefully when using blades and be sure to use equipment that has been kept in good condition. Master the following basic techniques, and then move on to create unique frames with a variety of shapes and sizes, colors, and textures.

materials required

two sheets of handmade paper
  (6 x 9in or larger)
cardboard
pressed flower
cotton paper for window
cutting blade
brayer
paste (see below)
brush

making paste

Place ½ cup of white flour into a small saucepan.

1

Slowly add ½ cup of cold water, whisking well to incorporate the flour as the water is being added. The mixture should be runny, like thin cream.

2

Place the saucepan over a medium-high heat and whisk well while slowly adding 2 cups of boiling water.

3

Continue whisking until the paste has thickened, this should take about 3–4 minutes. Let the mixture cool before using.

4

Cut two pieces of cardboard to 4 x 4½in. Use either bookbinding board or heavy cardboard. 1

Trace a window of 1¾ x 2¾in onto the center of one board. 2

Using a cutting blade, cut away window. 3

Cut a ½ x 3in piece of cardboard for the stand. Set aside. 4

Trace the back board onto one sheet of covering paper and trace the board with the window onto the other sheet. Here we are using Botanical PaperWorks® Moss paper made from abaca fiber. 5

tip

Allowing the boards to dry thoroughly before assembling keeps them from warping.

Apply a thin layer of paste to the back board and place on a sheet of the cover paper. As in the hardcover photo album project (see page 97–99), flip the sheet over and smooth in place with a brayer. Miter the corners and paste the edges. Put under a weight to dry.

6

On the cover paper with the window traced on it, make a diagonal cut from one corner of the window to the other. Make a second cut from the opposite corners, creating an "X" cut.

7

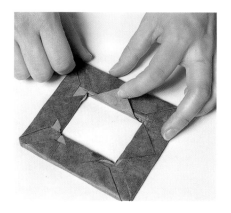

Apply paste to the window and place on the cover. Miter the corners and fold the edges over. Apply paste to the flaps of the "X" cut inside the window and fold over to cover the inside edges.

8

Apply paste to a scrap of cover paper and wrap around the cardboard stand.

9

Cut white cotton handmade paper to 2½ x 2½in. Center on the back board. Paste in place.

10

Apply a thin layer of adhesive to the back of the pressed flower, either spray adhesive or white glue. Center on the white paper.

11

Apply adhesive to the back of the window frame and position onto the frame back.

12

To make the object in the frame removable, glue the frame to the back on three sides only, leaving the fourth side open. Slide a picture into the open side.

13

Fold the cardboard stand at 1in and apply glue to the smaller portion. Position on the back of the frame. Press until dry.

14

tip

To create a magnetic photo frame, apply a small magnet to the back of the frame in place of the stand.

window greeting card

The window greeting card is a variation on the technique of making a photo frame (see pages 101–104. The card requires precise measuring and cutting skills, so take things slowly at first, and then move on to produce unique designs and embellishments. Use this technique to frame photos as greeting cards, or to create your own individual mini-watercolors to insert in each window.

1 Using a bone folder, fold one 6½ x 9½in sheet of handmade paper in half, to create a 4¾ x 6½in card.

2 On the front of the card, trace a rectangle measuring 1½ x 1½in, 1½in from the top of the card.

materials required

two sheets 6½ x 9½in handmade

paper, medium weight

small sheet of white handmade paper

white adhesive

bone folder

ruler

pencil

cutting blade

gold pen

petroleum jelly

Cut out the rectangle using a cutting blade.

3

Cut the second sheet of 6½ x 9½ in handmade paper in half. This second sheet of paper will be glued to the back of the card front, to hide the layered paper of the window.

4

Slip the cut sheet into the card and lightly trace the cut-out window. Remove the sheet from the card.

5

Cut a piece of white handmade paper to 3 x 3in. Apply adhesive to the back and place on the traced window, covering the pencil marks.

6

Apply a pressed flower to the center of the white paper.

7

On the front of the card, outline the cut-out window with a fine-tipped gold pen.

8

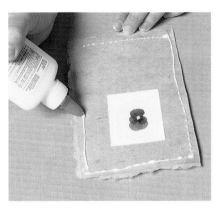

Apply adhesive to the outer edge of the cut sheet. Press in place inside the card, so that the flower shows through the window. Gently rub the card to secure.

9

embellished bowl

The papercast bowl can be both decorative and functional.
Collect interesting leaves and embellishments to begin building a collection of beautiful papercast
objects. This bowl is great as a display piece on a table or mounted on the wall. Make a series of these
bowls and mount them to form a feature wall display.

materials required

glass bowl	blotting cloth
damp, pressed sheets of cotton	petroleum jelly
handmade paper	clay beads
colored embroidery threads	fern leaf
three dried salal leaves clay beads	blunt knife

Apply a thin layer of petroleum jelly to
the inside of the glass bowl.

1

Pull several sheets of cotton paper.
Press but do not dry. Tear damp cotton
paper into small pieces and press into
a glass bowl. Layer them on top of
each other to cover the bottom of the
bowl.

2

Blot the paper with a cloth, to absorb
excess moisture and encourage the
pieces to bond together.

3

Position a salal leaf on the rim of the
bowl and cover with several small
pieces of damp paper to hold them in
place. Repeat with two more leaves.

4

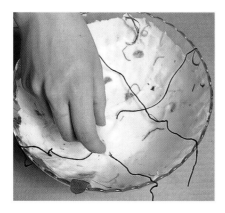

Lay embroidery threads across the bowl. Tear sheets of damp cotton paper, and press them into the bottom of the bowl to hold the threads in place.

5

Place a fresh or dried fern leaf on the bottom of the bowl. Cover with small sheets of cotton paper, enough to anchor them in place without completely covering the fern.

6

Place in front of a fan to dry. When completely dry, remove using a blunt knife.

7

To complete the project, embellish the bowl by tying clay beads onto the ends of the embroidery threads.

8

paper packages

Gift wrap has assumed huge visibility in today's market. Unique packaging for gifts has made the package itself an important part of the gift. Here you can learn how to make custom packages to enclose those special gift items. A beautiful envelope for enclosing money, a pillow box for small items, and a gift bag that can be adapted to any size, are wonderful ways to display your papermaking products and to add a very personal touch to your gift-giving. Simply scale the measurements to suit your particular needs. You could make several to keep on hand for quick packaging solutions, or even give the packages as gifts to be used by others.

materials required

two sheets 8½ x 11 handmade paper, same or different colors

two sheets 8½ x 11 plain white paper

white adhesive

ruler

cutting blade

bone folder

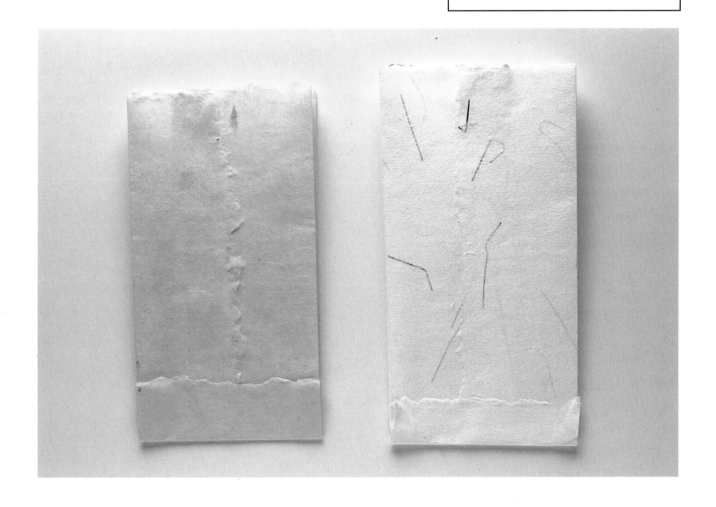

money envelope

When giving gifts of money, it is difficult to know how to present it in a way that expresses warmth and caring. A handmade paper envelope is the perfect way to give it that personal touch.

Create a paper template by measuring and drawing the dimensions shown onto a sheet of plain paper. Cut out the template using a blade and ruler. Apply a small dot of white adhesive, and position on the wrong side of the handmade paper. The adhesive will help hold the template in place.

1

Trace the template and trim the handmade paper. Remove the template and score as shown.

2

Fold the envelope along the score lines.

3

Secure the flaps in place by applying a thin layer of adhesive and smoothing in place.

4

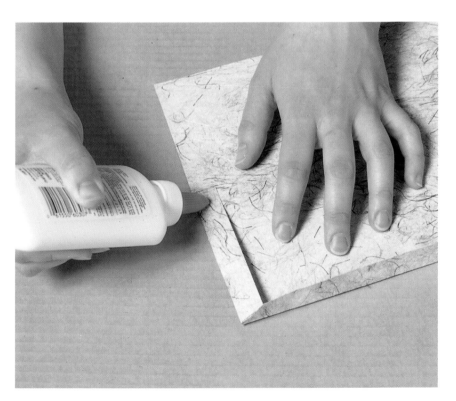

pillow boxes

These are the perfect package for smaller, fragile objects, because they can be surrounded in tissue and easily inserted into the pillow package. No extra wrapping is needed, since the pillow is made from handmade paper and is a gift in itself!

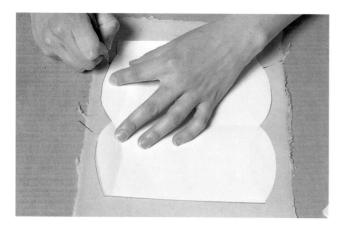

Create a paper template and trace it onto the handmade paper.

1

Cut out the handmade paper, and score along the lines as shown. Fold along the score lines.

2

Apply adhesive to the side fold and press in place. Pop out the end flaps to create a small pillow.

3

gift bag

Simple paper bags make wonderful containers for small gifts and favors for special occasions such as weddings and birthday parties. The bags are especially lovely when made from handmade paper that coordinates with your personal stationery or special-event invitations. Before using the handmade paper, practice these instructions on a piece of plain office paper in order to perfect the technique first.

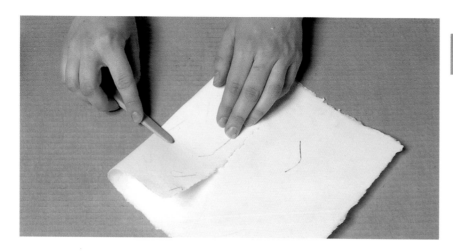

Place the sheet of paper on the table with the long edges at the top and bottom. Fold one side of the paper at $^{23}/_4$in.

1

materials required

one sheet 8$^1/_2$ x 11in handmade

paper

bone folder or blunt knife

adhesive

pencil

ruler

Turn the paper over and fold the other side at $^{31}/_2$ in. The two flaps should overlap by about 1in. Smooth the folds with a bone folder.

2

Unfold the sheet and mark $^1/_2$in on either side of the folds made in steps 1 and 2. These two measurements will create the side accordion folds of the bag.

3

Using a ruler a bone folder, score lines using the measurements in step 3.

4

With the bone folder, fold along the score lines to create an accordion.

5

Apply a thin line of adhesive to the side flap and smooth the other side flap on top. Press under a stack of books or press with a warm iron.

6

Mark a fold 1in from the bottom.

7

Cut away one side of the bottom flap in order to make the fold less bulky.

8

Fold the bottom flap over the cut edge and apply adhesive.

9

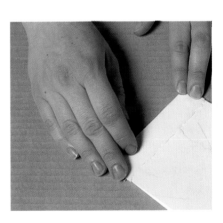

Press to secure.

10

sparkle paper

Fascinating sheets of paper containing a variety of light-catching sparkles can be created using readily available materials found around the home. Unique papers for any occasion can be made using the huge variety of metallic shapes and sizes available in party supply stores. Birthday party invitations can contain numbers representing the age of the guest of honor. New Year's party invites can sparkle with hats and noisemakers, all in multicolored foil cut-outs. Be on the lookout for sparkly bits, which you can store until the right moment!

materials required

Almost anything!

After beating and straining ½lb of cotton pulp, measure and add ¼ cup of metallic confetti. Blend the confetti and pulp together by hand.

1

Add the pulp to the vat as shown in Chapter 2 – Papermaking Techniques (pages 19–47), and pull the sheets. The inclusions can be seen in the pulp as some pieces rise to the surface of the sheet and others are incorporated into the pulp.

2

Couch the sheet on a prepared couching surface. Press and dry as desired.

3

Here is a selection of interesting inclusions, taken from supplies from around the home. Shown clockwise starting top right: snippets of fuzzy wool, pieces of yarn with metallic thread, chunks of gold foil, silver tinsel, and pieces of metallic tissue paper.

party crackers

Have you ever wished that you could personalize a set of party crackers to contain gifts of your choice, verses or words of wisdom, jokes, and party hats all specially chosen for each guest? Here is your opportunity to customize your party crackers for any occasion. Handmade party crackers are easy to make and a lot of fun for all ages. The crackers can be placed at the place setting of guests at a holiday dinner or handed out at a New Year's party. Imagine the fun, as your guests find a personally prepared surprise in each one!

Wrap the tube lengthwise in handmade paper.

1

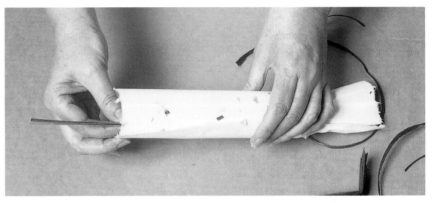

Insert the pull-snapper, so that the ends protrude equally from the each end of the tube.

2

tip

Party crackers can also be used as wedding favors. Correspond the paper that covers the cracker with your place cards or programs. Use satin or organza ribbon in place of raffia ties. Inside, include a message from the bride and groom, party mints, and a small puzzle or game.

materials required

one sheet of 8½ x 11in handmade paper
one cardboard toilet tissue tube
one pull-snapper (available at craft supply stores)
one party favor (small novelty erasers, bouncing balls, a wrapped chocolate etc.)
joke, fortune, or verse handwritten on a 1 x 3in strip of paper
one folded paper crown (instructions on page 119)
ribbon ties for each end

Tie one end with ribbon tie. Fill the tube from the other end with a party favor.

3

Insert a joke, verse, or fortune and a paper crown.

4

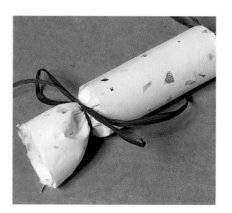

Secure the other end tightly with ribbon. A small line of glue will seal the seam if needed.

5

paper crowns

1 Cut the sheet into four strips, each measuring 5 x 30in.

2 Take one strip and fold into an accordion with folds of 2in.

3 While folded, cut a "V" across the top.

4 Open the strip, trim to 24in, and glue the ends together.

materials required

(makes 4 crowns):
one sheet of tissue paper 20 x 30in
glue or tape
scissors

PAPERMAKING

gallery

gallery

The author and publishers would like to thank the following people for contributing pictures of their work, to help show the beauty and versatility of paper as a medium.

Brenda Turner
(France)
Brenda has been making paper for several years. starting with a kit in the kitchen, and now works and teaches in her own paper making studios. Although the paper itself is an inspiration, her work is also much influenced by nature and the history and mystery of her environment.

Toni Smith
(Sydney, Australia)
Toni has been making paper for around six years, and is a member of Primrose Paperworks Co-operative Ltd in Sydney. She has exhibited (mainly artist's books) in Sydney, Queensland and Victoria.

Beverley Folkard
(Essex, UK)
All pieces made using Beverley's own paper, handmade from recycled sources.

Donna Allgaier-Lamberti
(Michigan, USA)
Donna is a writer, photographer, papermaker and artist, and works in a variety of mediums, both traditional and non-traditional. As a papermaker she is currently working with native plant fibers.

Geraldine Pomeroy
(Tasmania, Australia)

Ray Bliss Rich (New Hampshire, USA)
Ray is a full time artist who has been painting almost exclusively in the delicate Aumi-e style since 1980. After using Japanses papers for several years, Ray now makes his own.

Rikki Mitman
(Texas, USA)
All photos by Rick Wells.

Akua Lezli Hope (New York, USA)
Award-winning poet and writer, Akua has been paper and glass making for several years. She was papermaking artist-in-residence at the Women's Studio Workshop in 2001.

Catherine Nash
(Arizona, USA)
Catherine has had a love of plants since childhood, and creates handmade papers from a variety of natural media. She has taught and exhibited for many years.

M.J. Cole
(Texas, USA)
M.J Cole specializes in small runs of cotton rag, local plant fiber and other traditional papermaking fibers, as well as 3D objects from pulp.

Claudia Lee
(Tennessee, USA)

Background picture:
Toni Smith, *Piano spiral*.
Piano hinge book, paper made from various plant fiber papers. Cover is Wallaby Grass (Danthonia) paper with Tallow wood leaves.

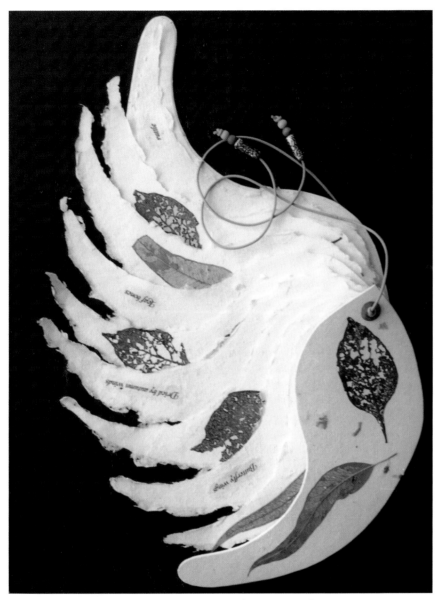

Toni Smith, *Leaf Bones. Fan book.* Cotton rag/cotton linter/recycled mountboard.

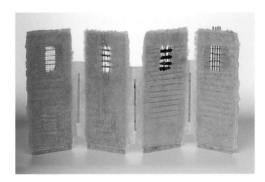

Claudia Lee. Table screen of handmade paper.

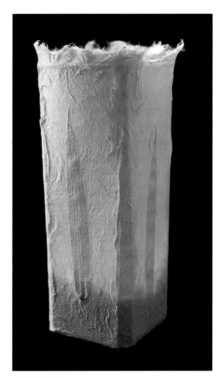

Rikki Mitman.

Brenda Turner, *From Flanders Field.* Hand made paper, sewn, torn and folded.

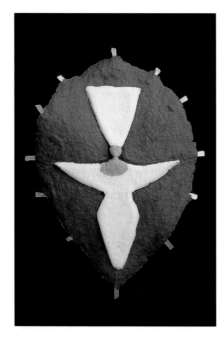

Brenda Turner, *Seedtime and Harvest.*
Hand made papers, vacuum formed
and pleated.

Donna Allgaier-Lamberti, *Spilling Open.*
Cast handmade bowl of corn husk, iris
abaca and cotton linter. Handpainted
tissue paper as embellishment.

Claudia Lee, *Rock, Paper, Scissors.*
Handmade and cast paper.

Rikki Mitman,

Donna Allgaier–Lamberti, *Boxes covered in handmade paper.*
Fibers include milkweed, goatsbeard, corn husks, iris kozo, abaca and cotton linter.

Brenda Turner, *Harvest.*

Akua Lezli Hope, *Radiant Protector Shield.*
Black denim, dyed cotton and abaca, copper, mold and vacuum formed pulp.

Catherine Nash, *Passage.*
Handmade paper of gampi bark and torch ginger grass.

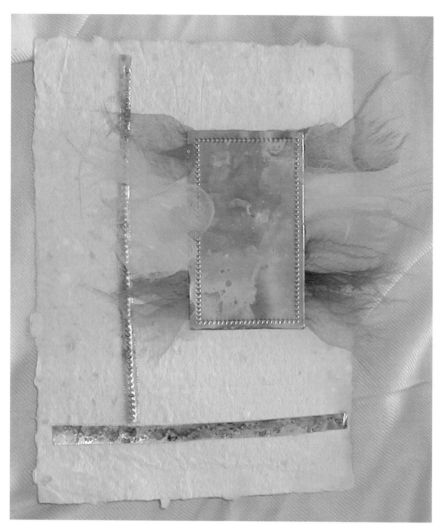

Akua Lezli, *Hope. Spirit House..*
Kiln-formed shield with kiln-cast glass,
tree branches, flameworked glass
beads, overbeaten flax fiber, copper,
copper wire and pacific stones.

Beverley Folkard, *Sampler.*

Beverley Folkard, *Sampler.*

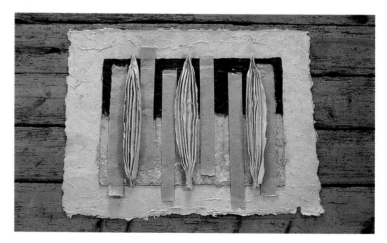

Catherine Nash, *Within Skin.*
A Passage. Environmental installation with large sheets of selfmade
Japanese paper of kozo (mulberry bark) and New Zealand flax.

M. J. Cole. Cotton rag pulp and plaster bowl. Surface finish is layers of colored gesso which has been sanded and handpainted.

M. J. Cole. Sheets of freshly made handmade paper and river birch bark were layered over lengths of string, then the string was pulled away creating fissures in the layers, showing the materials underneath.

Ray Bliss Rich, *Red Grass Flower.* Painted with Sumi ink on a paper circle, hand-pulled by the artist, showing its natural deckle edges. Collaged with winter killed knotweed paper (also hand pulled by the artist) on a background of Japanese chiri paper.

Geraldine Pomeroy, *Autumn leaves.*
Artist's book of loose leaf pages made from 100% cotton and fall leaves, cover decorated with scrapped fall twigs.

bibliograpy

- 300 Papermaking Recipes, Heidi Reimer-Epp and Mary Reimer, Quarto Publishing, Inc., 2000

- The Encyclopedia of Papermaking and Bookbinding, Heidi Reimer-Epp and Mary Reimer, Quarto Publishing, Inc., 2002

Creative Bookbinding by Pauline Johnson
Published in Canada by General Publishing Company, Ltd.
Copyright by University of Washington Press, 1963

Non Adhesive Binding Volume 1 by Keith A Smith
Published by keith smith BOOKS, March 1991 (1st printing)
Copyright Keith A. Smith, 1990, 1992, 1993

Non Adhesive Binding Volume 2 by Keith A Smith
Published by keith smith BOOKS, May 1995
Copyright Keith A. Smith, 1995

Hand-Made Books by Rob Shepherd
Published by Search Press Limited
Text Copyright Rob Shepherd, 1994
Photographs and diagrams copyright Search Press, 1994

Papermaking Techniques by Jean Kievlan
Copyright by Design Originals by Suzanne McNeill, 1995

Making and Decorating Your Own Paper by Kathy Blake & Bill Milne
Published by Sterling Publishing Company, Inc, 1994
Copyright Kathy Blake and Bill Milne

Cover to Cover by Shereen LaPlantz
Published by Lark Books, 1995
Copyright Shereen LaPlantz, 1995

The Art and Craft of Papermaking by Sophie Dawson
Published by Lark Books
Copyright Quarto Inc., 1992

Gorgeous Paper Gifts by Susan Carroll & Barbara E. Swanson
Published by Lark Books
Copyright Susan Carroll and Barbara E. Swanson, 2000

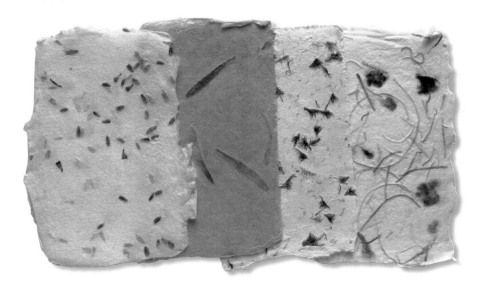

index

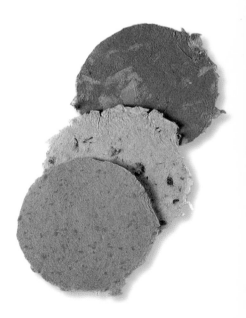

credits and acknowledgments

Thanks to Janet Carroll, marbler and bookbinder, for her expertise and contribution to the marbling section of this book. Thank you for sharing your techniques and designs.

Heartfelt thanks to Toni Reimer who cheerfully and capably took on many extra duties at BPW in order to free us to write this book.

Also to the staff at Botanical PaperWorks whose creativity and ability to make any occasion a party has created a workplace environment like none other!

To Alvina Pankratz for her endless enthusiasm for exploring and experimenting and her delight in discovering.

To Gary Reimer and John Reimer-Epp we give our love and thanks because they never cease to encourage and support our adventures.

Baby Ella who waited just long enough to be born to give Mom and Grandma time to finish the book!